ME, GOD
AND THE
MACHINE

ANN VAN WINKLE

D0109004

HARVEST HOUSE PUBLISHERS
EUGENE, OREGON 97402

ACKNOWLEDGMENTS

To my husband, Jerry, I give the medal of love for his understanding, patience, help, and encouragement.

My son, Jeff, receives my medal of joy. His support, encouragement, understanding, and his very being gave me the determination to live.

My family is due the medal of honor for their genuine concern, constant support, help, and love.

My friends receive the medal of commendation for their comfort, assistance, loyalty and spiritual uplifting through trying times.

My doctors, nurses, and technicians receive the gold medal of dedication for saving my life and the lives of thousands of kidney patients around the world.

And to God, my Creator and Savior, all glory and honor be given for preserving my life and for making it possible to tell my story.

Scripture verses are from *The Living Bible*, Copyright © 1971, Tyndale House Publishers, Wheaton, Illinois. Used by permission.

Some names of persons and places have been changed to protect the privacy of the individuals involved.

To Jerry and Jeff,
my loving husband
and
priceless son.

FOREWORD

For years there has been a strong research effort directed at substitution for various organs of the human body. The kidney became the first important organ for which an artificial substitute was used successfully.

In early 1960, Dr. Belding Scribner at the University of Washington Medical School pioneered the "team approach" to treating chronic end stage renal failure patients with an artificial kidney. This program was very successful in that the first five patients lived despite having lost their own kidneys.

However, in the State of Oregon, there were no funds available for treatment of end stage kidney patients. Some funds finally became available in the mid 1960s through the efforts of the Kidney Association of Oregon. Patients were screened and only those who could go into home dialysis were accepted. It was in this setting that Ann Van Winkle, one of the earliest patients in Oregon, developed kidney failure. She was reviewed by a medical screening committee and became a home dialysis patient. During this time, she had to endure many of the uncertainties and problems that patients have. Eventually she became a successful unrelated donor transplant. Her continued positive attitude and enthusiasm about life have made her an inspiration to know.

—**Richard F. Drake, M.D.**

CONTENTS

1   Three Choices   9

2   The Beginning   17

3   Please, Somebody Help!   25

4   The Training Field   35

5   Bye-Bye Salt Shaker   55

6   Wheelchair Days   61

7   Stepping Stones   75

8   Emergency!   93

9   A Way of Life 103

10   Kidney at Last? 115

11   Second Chance 135

12   A New Life 149

13   And Tomorrow Came! 157

These trials are only to test your faith, to see whether or not it is strong and pure. It is being tested as fire tests gold and purifies it—and your faith is far more precious to God than mere gold....

—1 Peter 1:7

Chapter 1

Three Choices

*I*t was four-twenty a.m. I was suddenly awakened by the warning alarm. Its buzzing sound
filled the bedroom.

I had fallen asleep just one short hour ago. At first
I thought I was dreaming. I sat up in bed. The pierc-
ing noise continued. My heart pounded as I reached
over to wake up my husband, Jerry, who was
sleeping soundly beside me.

No, I thought, let him sleep. Maybe I can do this
by myself.

I glanced toward the end of the bed and I could
see the red warning light flashing. Inside I felt sick.
I wanted to panic but something within me said,
"No, you can do it."

I leaned onto my elbow as the alarm continued to
pierce the night quiet.

My arm, I thought. It feels wet. It feels really wet.
What could be happening?

I glanced quickly at the drip bulb on my artificial
kidney dialysis machine. If everything was work-

ing properly, a steady stream of bright colored red blood continuously flowed from the tubes in my arm, the cannula, through the kidney machine, and back into my arm.

These tubes connected to an artery and a vein were my life lines. They lay exposed on the outside of my arm and carried my blood through the kidney machine to be cleansed. If these tubes were accidentally kinked or pulled out, the blood flow would cease and the alarm would ring. It was ringing now!

When I moved again, my hand felt a big wet spot.

"Honey, wake up! Wake up!" I cried. "I'm bleeding!"

In an instant Jerry was awake and by my side. As he flipped on the light, I knew we were in trouble. The bandage covering the tubes in my arm was soaked with blood.

"Get your bandage off!" Jerry exclaimed. "Maybe your tubes came apart!"

I began to shake as fear engulfed my whole body. This was the blood from *my* body soaking into the bedding and the mattress. The spot was growing larger by the second. I was afraid and was beginning to feel weak. I was losing blood rapidly.

For a fleeting instant my mind recalled sitting in Dr. Zee's office. As he finished my examination he looked me straight in the eye.

"Ann, you know I'm not God! I'm not sure what to tell you. These are all signs of kidney failure: the itching, the breaking out in a rash, the headaches, and the depression. There's nothing much we can do except put you on a kidney machine or perhaps put you on a waiting list for a kidney transplant. However, you should understand that these are both very trying situations—especially for the family. Kidney patients need much care and attention, and then, often times, they don't make it. Some people are actually better off dead!"

DEAD!? That word rang through my head like a deafening alarm and I didn't hear another word he said.

"No! Not death for me!" I kept saying over and over again in my mind.

But now, I thought, maybe he was right. Maybe I would have been better off dead. Maybe this is the end right now.

As I jerked the rest of the bandage off, blood spurted out from underneath.

"It's an artery!" I cried.

"Just stay calm," Jerry said. "I know what to do."

He reached for the blood pressure cuff and used it as a tourniquet around my arm. As he pumped it up, it became tighter and tighter.

"It hurts!" I cried.

"It has to be that tight to stop the blood flow," he said. "I'll release it every few minutes. Now where is the doctor's number?"

"Right there on the phone," I answered.

While Jerry was on the phone, I laid my head back on the pillow and tried to tell myself everything was going to be all right. But I felt as if I were going to faint, so I closed my eyes. I could hear Jerry in the background.

"Okay, we're on our way. We'll meet you in the emergency room."

"Call Susanna," I cried. "I know she'll come over and take care of Jeff until your mom can get here."

Jeff, our two-year-old son, was asleep in the next bedroom. Fortunately, he had not awakened during all of the commotion.

Before five minutes had passed, my husband had everything under control. We had been trained in emergency procedures at Good Samaritan Hospital in Portland, Oregon. They had also trained us in all the procedures necessary to run a kidney machine, and Jerry had learned his lessons well.

I heard someone at the front door now. It was Susanna, our neighbor who lived across the street.

"Just stay with Jeff until Grandma gets here," Jerry said. "She should be here very soon. I'll carry Ann out to the car if you'll just hold the door open for me."

"Sure will," she replied. "You'll be all right now. I just know it. Don't worry."

My strong husband grabbed several towels and me at the same time and carried me out to the car.

"You're going to be all right," he assured me. "Dr. Hayes will meet us at the emergency room. He said to bring you in the car. It will be faster than waiting for an ambulance."

Dr. Hayes was my surgeon, and I had confidence in anything he said or did.

The trip in the dark of night was frightening. I felt panic inside again and I laid my head over on Jerry's shoulder. I knew he would take care of me, but I worried we would not make it in time. My arm felt numb and I knew any moment might be my last.

There were no policemen in sight as we drove toward Portland. "Wouldn't you know we wouldn't find one when we need one," Jerry joked. "One lucky thing, though—at this time of night, there's no traffic. We're only about fifteen minutes away. Just try to think of something else."

As the car sped on, my mind drifted and recalled the few short months ago, at only twenty-seven years old, when doctors had given me but three choices: have a kidney transplant, go on an artificial kidney dialysis machine, or die!

Our marriage was a short seven years old, and our only child, Jeff, was bursting with life and all the goodness of many wonderful years ahead.

"God, please help me," I thought.

No, I would not die! Life, despite the many perplexing problems we had experienced so far, was too good, too exciting, and too adventurous to give up now. Seeing my child grow up became my number one priority.

Another option available to me was a transplant to replace my own weakened kidney that had been failing for at least a year and a half. A kidney transplant was a remote possibility, but this delicate operation was only in the infant stage in the wonderful world of medicine in 1967. True, we lived very close to Portland, Oregon, and the University of Oregon Medical School (since renamed The Oregon Health Sciences University) where they had been doing a few transplants since 1958. These tranplants, however, were performed mostly on patients who had relatives willing and qualified to donate a kidney. They preferred the donor to be a twin brother or sister. I had neither. We also knew that the lucky kidney recipient had to be in good health, and at this time, my wasted, sickly body was in no condition to undergo surgery. A kidney transplant would not be a wise decision right now.

We decided that the kidney machine was the best alternative, despite Dr. Zee, who said I would be better off dead. He just didn't believe in people living on machines!

These recent memories vanished for a moment as I glanced at my arm. My fingers had no feeling and my body felt weak and limp.

"I think this blood pressure cuff needs to

be loosened, honey," I said.

"It will be okay. We're almost there now."

Again, I felt a surge of panic set in. I just knew I was going to faint or die. Maybe I didn't even care. I thought about life on a kidney machine.

I had to be connected to the machine three times a week for ten hours at a time. Family life was stressful. There were no kidney centers to go to so Jerry had to run me on the machine at home—away from the all-knowing, watchful eyes of doctors and nurses. It was expensive, uncomfortable, and frightening.

As I was questioning my thoughts, we pulled into the emergency parking space at the hospital.

"Stay here. I'll go get the nurse," Jerry calmly said.

It was only a few seconds, but it seemed like hours before Jerry and the nurse appeared with a wheelchair.

After examining my arm, the nurse wheeled me directly up to surgery. Dr. Hayes met us and he was a wonderful, reassuring sight. He had surgically installed my cannula tubing a few weeks ago, and I had great faith in him. When he was near, I knew everything would be all right.

The operation to re-connect the cannula tubing in my arm lasted about an hour and a half.

"You'll be as good as new!" Dr. Hayes said. "You handled that emergency very well. Now, I just want you to stay in the hospital overnight and we'll see how your arm looks in the morning. I think you'll be able to go home then. The nurse will wheel you down to your room and you just get some good sleep. I'll see you in the morning."

"Thank you, Dr. Hayes," I replied.

As I gratefully lapsed into sleep, I knew I would

be facing other problems tomorrow. In fact, it would not be long before doctors would predict I would never walk again and that my life expectancy would be a short two to five years more.

Trust in the Lord with all your heart and lean not on your own understanding.

—Proverbs 3:5 (NIV)

Chapter 2

The Beginning

*I*t all began with a simple urinary tract infection, just six weeks after we brought our new son, Jeff, home from the hospital. At last, after having lost two babies, we really had a healthy, beautiful son. We were overjoyed and gave thanks to God.

The trouble started one night after dinner. It was a wintry November evening in 1965. I hadn't been feeling well for the past few days, and blamed it on being a new mother. Having a new baby always demands more attention and energy. There were feeding schedules to maintain, diapers to change, crying to endure, more washing, and also the house and meals to care for.

Jerry had been good to help, but still there were sleepless nights for me, and maybe all of this was just getting me down.

I had picked at my dinner this particular evening and finally gave up and pushed back my plate.

"What's wrong?" Jerry asked.

"I don't know," I replied and broke out into tears of exhaustion. "Maybe I'm just not cut out to be a mother. I'm so tired and depressed. I just don't feel very well."

He moved over and put a reassuring arm around my shoulder.

"Don't worry," he said. "Everything will be all right. Why don't you just go to bed, honey? It might be a good idea to take your temperature first. I'll take care of the dishes and Jeff. We will get along just fine. In fact, we're going to talk about hunting bears—teddy bears, that is! We think Mom needs a good night's sleep, don't we, partner?" Jerry turned to get little Jeff's approval.

Feeling somewhat reluctant, I left the dishes for the boys and headed for the bathroom to get the thermometer. As I searched through the untidy bathroom drawer for it, I briefly wished I were a better housekeeper.

"Oh, well," I quipped, "never do today what you can put off until tomorrow!" Housekeeping at this point was the least important thing on my mind.

At last! My fingers rolled across the round thermometer that had fallen precariously to the very back of the drawer.

I headed for the bedroom. My body felt weak and I broke out in an uncontrollable sweat.

"Boy," I thought as I flopped my wavering body onto the bed. "I really don't feel well!" Sticking the thermometer under my dry, swollen tongue, I closed my eyes.

After a few moments I reached for the thermometer. The mercury was difficult to read. I had to raise my heavy head and open my eyes wider.

It's going up, I thought, and I shook my head to look again. It read one hundred degrees.

Oh, dear, I thought. What now? I certainly hope I don't have the flu. That's one of the last encounters I need with a new baby!

I disgustedly put the thermometer back into my mouth.

"Leave it for at least five minutes," my mother had always said. My childhood experiences with many cases of strep throat had etched Mother's bit of advice into my being. Her meticulous care for her children (especially sick children) had always been right.

I closed my eyes again. The pattern of shadows on the bedroom wall remained in my vision. Lights from the neighbor's house and trees in the backyard cast feverish figures on the bedroom wall. It had been a cozy bedroom, but now the sweet fragrance from the honeysuckle vine that draped the window was nauseating.

We had been lucky to find this house and rent it. It was just an empty house—without a "For Rent" sign—that we just happened to spot on a drive one afternoon.

This house was surely a gift from God and at this time, we didn't know how true this was. Unbeknown to us, across the back fence lived Charlie Willock, the inventor of the new home dialysis artificial kidney machine.

God knew what He was doing when He placed us in this chosen home long before my illness was revealed. Mr. Willock would play an important part, a very important part, in saving my life in the months to come. We still had much growing and trusting to do, however, before God's total plan for our lives would be even partially revealed.

I released my cramped mouth from the thermometer for the second time. One hundred and three degrees! No wonder I felt terrible! I'd call the doctor in the morning.

I probably just have the flu, I thought. Right then my back hurt, I had the chills, and all I wanted to do was to go to sleep.

It was a restless night. I vaguely remember Jerry crawling into bed. The rest of the night turned out to be a nightmare, filled with restless dreams and feverish visions.

The next morning I called my obstetrician, Dr. Franel. I had great faith in him, for he had brought me through a successful pregnancy. He also reminded me of my father, whom I loved very much which was comforting. Daddy was always positive, always reassuring, and always knew what to do. I knew Dr. Franel would also be helpful, and I picked up the phone to call him.

The receptionist answered.

At my request to talk to Dr. Franel she told me he was with a patient, but would return my call shortly. It wasn't long before he called and asked how I felt.

"I'm not feeling very well this morning. My back hurts, my temperature goes way up to one hundred and three degrees and then drops to below normal," I answered.

"Do you have any burning when you urinate?" he asked.

"No, I don't think so," I replied.

"It sounds like a mild kidney infection," he answered. "I'll put you on some sulpha drugs and we'll get it cleared up in no time!"

"I hope so," I answered. "I'm trying to get back in shape so I can go back to teaching school. I have a contract to start again in just a few weeks."

"Don't worry," he replied. "We'll get you fixed up."

However, this conversation became an all too common occurrence. A few weeks after I finished the medicine, my fevers would return. Another kidney infection!

"We'd better put you on some stronger medicine," the doctor would say. And after four months of recurring infections, he thought it best to send me to a urologist. My new doctor decided to admit me to the hospital to do more extensive urinary tests.

So we arrived at the hospital in the late afternoon. Like most people, hospitals never were my milieu. I could think of many other places I'd rather visit or stay. Walking through the front door was like walking into a huge medicine chest. Smells particular only to these institutions permeated the very walls that surrounded us.

We registered with the front desk personnel, answered many questions, filled out time-consuming forms that indicated our financial responsibility, and then took the elevator to the fourth floor—my assigned residence. We walked into the middle of an emergency.

Nurses and doctors hovered over a body that lay sprawled out on a gurney in the middle of the hallway.

"Did you call for the machine?" a doctor asked.

"Yes, Doctor. It's on its way," a stately nurse answered.

At this same moment, another elevator door opened and the attendant quickly maneuvered a large machine out of the doors and onto the floor.

"Here's the machine now, Doctor."

"Good. Prepare the patient for a cut down. We need to get him on dialysis as soon as possible."

As we stood there observing the action, we learned that this awkward-looking machine was an artificial hemodialysis machine (commonly called a kidney machine). The man whose life was apparently in question had overdosed on drugs and his kidneys had quit working. He would need to have these poisons cleaned out of his bloodstream if he were to live. Life was the goal here and everyone was doing his best to win.

We reported in at another desk on this same floor and a nurse showed us to my room. She took my blood pressure and my temperature, then left me a hospital gown, saying the doctor would be in soon.

As I undressed and then climbed into my hospital bed, I felt a little shaken by the emergency we had witnessed in the hallway.

I didn't know much about kidney machines, but they didn't look like fun. I recalled an experience I had while attending Portland State College. I had driven our car to school one sunshiny morning, happy about the day and classes I was attending. Just after I parked the car, I reached to turn off the radio and was captured by an announcer's plea:

> Gloria Brown of Longview, Washington, needs your help. She is dying of kidney failure and needs the treatment of a kidney machine. With your help, she will live. Please send one dollar, five dollars, or one hundred dollars to Gloria Brown Foundation, P.O. Box 000, Longview, Washington. That address again is P.O. Box 000, Longview, Washington. Thank you for your help and support.

"Why not?" I had said to myself. "It must be terrible knowing you could have the chance to live if only you had enough money."

I took out my checkbook, wrote a check for five dollars, and sent it to Gloria. It was a wonderful feeling to know I had helped someone in need.

My train of thought was broken by the arrival of my doctor. He explained about the tests they would be doing on my body tomorrow and reassured me not to worry. Nothing was going to hurt because I would be asleep.

The next day came, bringing with it many personal apprehensions. I survived all of the tests and late that afternoon the doctor came with his diagnosis.

"You don't have anything more than a kidney infection as hard to get rid of as dandelions," he quipped. "We'll just put you on some stronger medication and send you home tomorrow. You'll be feeling fine in a few days. Don't worry, we'll have you all fixed up very soon."

I believed him! At last—someone knew what he was talking about. A new doctor, new medication—this was the ticket! We had finally found help.

Our bubble was burst all too soon, however. After taking the medication for ten days, I felt great. Then, all of a sudden, "friend fever" came to stay again.

I was back teaching now, and time off from a new job was difficult. Sick leave did not last forever.

"We'll give you another dose of stronger medication," the doctor announced. "This will do it. I'm sure you'll be fine in a few days."

More medication ran into more frustration. Headaches became very frequent along with puffy eyelids, fingers, and ankles. My backbone itched uncomfortably, and depression set in without budging.

Something more had to be done so we decided to change doctors again. This time, Dr. Zee was our choice. My father had gone to this prominent urologist in the city. He surely would know what to do. We proceeded with hope in our hearts.

God is our refuge and strength, a tested help in times of trouble.

—Psalm 46:1

Chapter 3

Please, Somebody Help!

*D*r. Zee was all business—that is, in everything except scheduling patients. After waiting three hours in the outer office to see him, the nurse finally called my name. She was very apologetic about the long delay, and I brushed it off casually.

He must be very good, I thought. Having to wait so long to see him must mean many people sought his care. The time he spent with them was no doubt necessary and I knew I would get my fair share of time.

The nurse took down all of my previous medical history and showed me to a nearby cubicle.

"Take off everything except your panties," she glibly instructed. "Put on this gown and tie it in the back. The doctor will be right with you."

These instructions, which I was used to by now, always guaranteed a race within me. What if the doctor entered and I was standing there in only my panties. How embarrassing that would be! Somehow, standing up without any clothes on was far more degrading than lying down on the slender, paper-covered table and having the doctor lift the gown exposing everything. I couldn't figure out the logic.

Accordingly, I raced to get my clothes off, and the zipper on my pants stuck fast! There I was, braless, feeling like a topless queen on the Vegas strip, when there was a slight tap on the door. My heart leaped as the door opened.

"Are you ready?" a voice inquired.

Thank goodness it was the nurse.

"Just about," I answered, "but my zipper is stuck. Oh, there it goes. I'll be ready in just a jiffy."

I was just tying the last string securing the air-conditioned back of my gown when in walked Dr. Zee.

"Hello, Mrs. Van Winkle? I'm Dr. Zee."

He spoke with a voice of authority as he continued reading my chart—almost unaware of my presence.

"Looks like you're having some infection problems." He gave me a quick glance. "Have you ever had kidney trouble before?"

"No," I replied.

"Any history of it in your family?"

"Well, my father has had a few kidney stones removed, by you, as a matter of fact, but that's all," I replied.

"Oh, yes, I do remember the name now. I see here you've had a history of strep throat."

"Yes, as a child I can remember having one case of strep throat after another. Our family physician always gave me medication for them, but it was never long before I had another sore throat."

"Strep has a tendency to go from the throat to the kidneys," he added. "We're always suspicious when urinary problems occur after a history of strep throats. Now I understand you just had a baby not too long ago. How's the baby doing?" he asked.

"Oh, just fine," I replied. "He's a healthy little boy."

"That's wonderful. Now, about you," he continued. His voice softened just slightly. "I'd like to admit you into the hospital and do a few routine blood tests and a needle biopsy of your kidneys. We may have to do a full kidney biopsy, which is major surgery, but at this point, I don't know.

"Now, plan to be in the hospital two or three days. The sooner we do this the better. This is Friday. How about Monday? Is that a good time for you?" he finished.

"I think so," I answered. "I'll just have to find someone to take care of the baby."

In the back of my mind, I thought maybe my sister-in-law would take care of Jeff. He would have a good time with his little cousins.

Everything was set up and I entered the hospital as planned. When morning came for the procedures, I was apprehensive as they wheeled me away to the surgery room. The doors opened automatically and inside lay a cold, sterile world. Everyone was dressed in baggy green uniforms, green hats, and stretchy green shoe cover-ups. Cardboard masks surrounded each one's mouth, leaving only peering eyes as individual characteristics.

More questions were asked, more forms filled out, and answers were routinely recorded. Then, the move onto the cold surgery table, an adjustment of the overhead light, a stick with a needle, and I was asleep before I could count to ten.

When I awoke, I was back in my hospital room. Dr. Zee and Jerry were talking.

"We did a full-scale kidney biopsy," he said. "She has an incision twelve to fourteen inches long around her right side. It will be sore.

"I have to tell you we discovered unpleasant findings. Your wife has *pyelonephritis* and *glomerulonephritis* of the kidneys. This means the top part of the kidney is infected and also the *glomeruli*, which are the filtering tubes in the organ. The tubes have been plugged up with scar tissue, and, because the kidneys do not rejuvenate themselves, they will never unplug. I will put her on a special salt-free diet, have her drink over a gallon of water a day, and give her more medication. If this doesn't work, well, the future may not be very bright. She may have to go on a kidney machine, have a kidney transplant, or die."

We couldn't believe what we were hearing. This couldn't be happening to us! We were newlyweds of only seven short years. We had a new little baby, Jerry was just graduating from college, and our lives were finally blooming. What we were hearing was something that happens to somebody down the street, or, better yet, some person who lives way across the United States in a remote town we have never heard of! This wasn't going to happen to us! It couldn't, I kept telling myself over and over again.

Time marched regimently on, and in the days following my hospital stay, we did everything the doctor told us to do, right down to drinking the last drop of one gallon of water per day.

Circumstances got worse, much worse, before they even started to get better. The frequent trips we made in to see the doctor often left us sitting in the waiting room four and five hours before we were called in to see him. Our total time with him, after waiting that long, would only be two or three minutes.

"Come back in a week," he would say. "Keep drinking your water."

I was getting very discouraged because it seemed like he wasn't doing anything for me, and it also seemed like he didn't even care.

I found myself getting sicker and weaker. I just didn't feel like myself and felt incompetent to teach school or take care of our child. Ladies from our church volunteered to help with meals and the baby, so we humbly accepted their charity. One kind, compassionate woman from the church, Aunt Lettie, was a special joy to me every time she came. She was so kind and loving to both little Jeff and me that we savored her visits.

Behind the scenes, Aunt Lettie was busy following God's guidance for our lives, though we did not even know it. She was an angel sent by God. Aunt Lettie knew Charlie Willock, who lived behind us and had, miraculously enough, invented the home dialysis artificial kidney machine! She told Charlie about me, and he began working on a plan to get me to see Dr. Drake, the doctor who helped him with his new invention.

These two men had met through their daughters who went to the same school. The Willocks invited the Drakes over to their house one day for an afternoon of swimming and the two men began exchanging some friendly conversation. Dr. Drake told Charlie about one of his patients who was dying and needed a kidney machine. Machines were not yet available, and Charlie, being an inventor, decided to try to make a machine for this patient, Mark Smith, who lived in Portland. The machine was developed in Charlie's garage, which was right over our back fence. This seemingly small invention was extremely successful and grew into a multimillion-dollar, worldwide business. Charlie Willock not only saved my life and the life of Mark Smith, but thousands of other good and useful lives through the world. What a blessing they and their machine were!

Many other people also worked among and between the professional strategies to save my life. I, however, was getting weaker and weaker. Finally it became necessary for me to spend most of my days in bed. I was so sick we decided it would be best to take me out to the family farm. My mom and dad would be in and out during the day and my grandmother would be able to come over and sit with me also.

Meanwhile, life, in itself, was getting extremely complicated and depressing. But there were some small pleasures, like having someone rub my feet. They were itching terribly all of the time now, practically driving me crazy. They were beginning to go numb and this never-ending sensation didn't go away.

We didn't understand everything that was happening, and didn't know what to do, but we know now that God was busy working through other people. It wasn't long before my doctor, Dr. Zee, had a change of heart. Up to this point, I sincerely believed he was letting me die. He thought that putting a person on a machine was too hard on the family.

Finally, though, when Jerry called him on the telephone to report on my condition, Dr. Zee decided to take me to Providence Hospital and try dialyzing my worn, sickly, degenerated body.

"Now, there's no guarantees with this," he said. "She may do fine and then again—well—"

"I understand," Jerry answered. "Anything would be better than how she is now. She is itching all over her body, can't sleep, has terrible headaches, and her feet and now her fingertips are going numb. My wife has played the piano for years, but her hand muscles are so weak now that she cannot even press down a key on the piano, and feeding herself is getting to be very difficult."

It was agreed upon that I would be dialyzed soon. The very next day we made the eventful journey to Providence Hospital in Portland. I checked into the hospital as a matter of routine; however, this was really a life-and-death situation and we knew it.

Death hovered too near, for when I was assigned to my hospital room, I found my roommate was an elderly woman in her eighties, who was not expected to live. Was I placed number two in death's row? I wondered.

When the hour came for the dialysis, the doctors did not put regular cannula tubing into my arm. Instead, they did what is medically known as a "cut down." Two small slits were made in my groin with an arterial tube hooked up to an artery in my leg and a venous line hooked up to a vein in the other leg. It was no picnic, and I was glad when they finally had me hooked up to this strange, frightening machine. All I understood was that the procedure I was about to endure would probably make me feel better, and I knew that I had to lie completely still for about four to five hours.

I remember wondering again—"Why is all of this happening to me? What does the future hold for us? Will I ever see my baby grow up? Why? Why? Why?"

Dialysis was new to Oregon in the year of 1967. It was used in the service during the war back in the forties, and the Veteran's Hospital in Portland continued a small dialysis program. In addition, the

State of Washington was pioneering a dialysis program on the West Coast at one of their leading hospitals, but their good fortune had not yet spread to Oregon.

Besides, money was not readily available. The Kidney Association of Oregon was just being formed and many questions lay unanswered— questions like: who was worthy of the expensive life on a kidney machine, and would people need to be sorted out like old silver coins to see what their worth to society would be?

Psychologists, psychiatrists, and other doctors began wearing a path to my bedside. They questioned my worth. I didn't realize at the time the judgments that were being made on my life. Thank God I was young, a mother of a small child, married, a teacher, and psychologically sound of mind. Otherwise, I may have exited—stage left!

When my first dialysis experience was over, I had to stay in the hospital for a couple of days and then I was released to go home. There was much to be done that I wanted to be a part of. Jerry was finally graduating from college and there was much preparation to help his mother plan a reception for after the commencement. I hoped I would feel well enough to help.

After the graduation, it was not long before the poisons resulting from lack of kidney function began rapidly building up in my system again, and my body and mind were in misery all over again.

On July 4, 1967, thirty-four days after my first dialysis experience, I was again admitted to Providence Hospital for more hemodialysis. This time I was even weaker and sicker than the time before, and no one but God knew if I would survive.

No children were allowed to visit in the hospital,

but a sister at Providence, Sister Joan, helped Jerry smuggle little Jeff in to his "mommy." What a joy it was to have my little guy crawl around on the bed saying "Mama, Mama." This was better medicine for my body and soul than all of the pills or dialysis in the world. It gave me that beautiful desire to keep going, to keep pushing, to keep striving to live because life was really worth it. My son needed his mother.

After the second dialysis was completed, things didn't look very good. My system had gone into a type of shock that sometimes occurs after a traumatic experience, and the doctor approached Jerry with dim hope for my life.

"Mr. Van Winkle," he said, "things don't look good. Not good at all. If I were in your shoes, I would call the family in and inform them that your wife may not live through the night!"

"Oh, my God," Jerry replied. "Not my wife!"

"I'm sorry," the doctor replied. "We're doing all we can at this point."

Jerry called the family and I know a lot of prayers were said for me that night. The power of prayer is a mystery not easily understood. God, in his infinite mercy and love, saw fit to spare my earthly life during those late night hours, and time marched on.

The very next day as I was lying tired and weak in my hospital bed, something happened. I became very irrational. My mind was extremely confused. Nurses just outside my door appeared to be going up and down the hallway hooked to trolley car tracks. I hollered for help. No one listened. No one heard. No one came.

I thought the hospital was a mental institution and I had to get out of there at any cost. The nurses

ended up putting me in a straitjacket. I don't remember too much more about the unpleasant experience; it is marvelous how God puts a shield over
our memory when necessary.

It was about ten days later that things began improving and I could once again go home.

During this time, many outside forces were still
working to get me in to see Dr. Richard Drake, the
head of the new home dialysis program at Good
Samaritan Hospital. The appointment was made
through Dr. Porter at the Medical School. This doctor had gone to lunch with Jerry and a mutual friend
several weeks before and had advised him to get
another doctor's opinion. Now was definitely the
time for this second opinion.

At this point, I was too weak to walk, and too
weak to even get dressed, so Jerry took me to Dr.
Drake's office in a wheelchair. I struggled into my
blue quilted robe and some bedroom slippers and
we were on our way.

Chapter 4

The Training Field

When we arrived at Dr. Drake's office, we were taken in to see him almost immediately. I am sure it was obvious how sick I was, and that any wait would have been much more than I could physically tolerate.

"May we start at the beginning?" Dr. Drake asked after we had met and briefly explained my plight. "When did you start getting ill, Ann?"

As I related events to him that had happened during my lifetime and specifically over the past year and a half, he took many notes.

"It is obvious that you need to be on a dialysis program," he said. "We have such a program here at Good Samaritan at the present time, but it requires a lot of cooperation from both the patient and the family. Jerry, that means you or some other close family member. It is a home dialysis program. We train you here in the hospital in all the procedures necessary to run an artificial kidney dialysis machine. When you have trained and run on it here in the hospital for eight to ten weeks, we then send you home and you run on the machine at home. Usually we require you to run three times a week for ten to twelve hours each time.

"Now this may sound difficult to you, but we have successfully trained eight people to do this procedure and they are now living useful lives and running on their machines at home. We always have someone on call at the hospital in case you have questions during your runs or have complications. You just learn how to deal with them mostly by yourself, but we are always here to back you up. Now, do you have any questions?"

"Well," Jerry said, "how expensive is this machine. We don't have much money. I've just graduated from college and we haven't had a chance to save up much yet. We have a one-year-old son, too."

"I don't have anything to do with the financial end," said the doctor. "But I do understand that Ann has been screened by a number of doctors. She may not realize it, but she has been accepted by the board of directors to be helped financially by the Kidney Association of Oregon. This organization has been in existence for only three months and is directed by Chuck Foster. I'm sure he can answer any questions you might have about financial aid. Now, Jerry, are you willing to learn to run this machine for Ann?"

"Sure," he replied. "When do we start?"

"The first procedure is to get a cannula put into her arm. We will surgically insert a tube into an artery and then another one into a vein in her arm. This tube will be our access to her blood supply.

When she is not being dialyzed, the tubing will hook together and the blood will flow freely through it. It will be partially on the outside of her arm and it will have to be kept covered by sterile dressings at all times. You'll have to take some medicine, Ann, called Coumadin so that your blood is less apt to clot. This tubing will protrude a little from the sterile dressing you will wear over it. You will need to check it often to see that the blood is a bright red color. This will mean that it is flowing freely through the tubing and is not clotted. If it clots, then we have a procedure we go through to unclot it. Also, Ann, at all times you will wear clamps on a chain around your neck to clamp this tubing off when it is unhooked to go onto the machine and to clamp it off if it ever separates accidentally. You could bleed to death very quickly if it separated and you did not clamp it off immediately.

"Now, we need to set up an appointment with Dr. Hayes for the cannula installation. Excuse me, please. I'll be right back."

As Dr. Drake left the room, Jerry and I looked at each other, both feeling relief and happiness, hoping that things were finally going to get better. It wouldn't be easy, but we could handle it with God's help.

My appointment with Dr. Hayes, the surgeon, was made for the very next day. I liked Dr. Hayes extremely well from the first time I met him. He was a large man, like my father, quiet spoken, and he had a warmth which radiated how much he really cared. I found out later that this beautiful, dedicated doctor came to regard his patients as personal friends. He became increasingly special to me, for in later months he was to save my life two times.

How fortunate we are that God "does not give us any more than we can bear." Jerry, Jeff, and I continued one step at a time to begin living again.

I was admitted to the hospital to have the cannula apparatus surgically installed into my arm. The surgery would be performed this same day—probably in the evening, Dr. Hayes had stated. He explained that I would not be put to sleep for this operation. The normal procedure is to just deaden the arm. He said I would be able to feel him poking around in there, but that there would be no pain. I trusted him from the beginning.

As I was wheeled down the hall on the gurney, I was a bit apprehensive, but thankful that I was finally getting some help. My thin, weak, miserable body couldn't have taken many more of the poisons that were building up from lack of kidney function. These poisons, we learned from Dr. Drake, had started attacking the nerves in my toes and fingers. He hoped they could reverse the destroying process. If it wasn't stopped, the muscles would be affected. He explained that the brain sends a message down the nerve, telling the muscle to move the toe or leg, etc. If that nerve were damaged, it would not get the message to the muscle and therefore, it would not respond. The result would be paralysis of that particular part of the body. Dialysis was the only way to reverse this process. In fact, if I had been dialyzed regularly and at the proper time by Dr. Zee, I would not have suffered this nerve and muscle damage.

As I lay there on the gurney in the hallway near the operating room, I remember repeating the Twenty-third Psalm over and over to myself as I had done on other occasions. Finally, a friendly nurse came to wheel me into the operating room. She wheeled me right up next to the operating table.

"Can you scoot yourself over onto this table?" she asked.

"I think so."

As I scooted my body over, I could feel the coolness of the surrounding room. I covered myself sufficiently and settled down on the operating table to await the next procedure. My eyes surveyed the room from my horizontal position. Above my head was a large, round light that was suspended from the ceiling by a long, moveable steel arm. Across the room and to my left were glassed-in cabinets containing rows and rows of medical supplies. There were two moveable stools, a waste basket, a telephone on one wall and a cart beside the operating table. The two nurses in the room were covered from head to toe in the usual green, hospital-type, baggy gowns. Distinguishing them from one another was not an easy task. I had to note the eyes carefully, for they were the single exposed feature.

One nurse approached me as she put on some sterile gloves.

"Hi, Mrs. Van Winkle," she said. "My name is Tanya and I will be scrubbing your arm with a Betadine solution in just a few minutes. We need to make the area as clean as possible. Now, how are you doing? Are you warm enough? We'll be covering you with some more drapes as soon as Dr. Hayes arrives."

"I'm just fine," I replied.

"Okay, now will you just hold your arm up in the air for me? This solution might feel a little cold at first, but it won't sting or anything," she said.

I said, "Okay," and she began dipping gauze into a brownish-colored solution and really scrubbed my right arm with it.

"Now, Ann, please keep your arm up in the air for a minute and I will dry it off with a sterile towel. Then we'll rest it back down on another sterile towel."

"Okay," I replied again.

"Perfect," she said.

About that time, Dr. Hayes entered the surgery room and came over to me.

"How are you doing?" he asked.

"I'm okay," I replied, "especially since you are here."

He emitted a feeling of peace, and I knew everything would be all right.

The nurses covered me from head to toe with sterile drapes, leaving only an opening to breathe through and an opening where my arm protruded. The doctor was ready to begin.

"Now, there's going to be a little stick," said Dr. Hayes. He had inserted a small needle into a nerve and injected an anesthetic that rapidly began to deaden my arm.

"Can you feel that?" he asked.

"Well, just a little," I replied. "It kind of feels like you're pressing on som thing."

"That's good," he ar swered.

The operation was tarted by cutting down toward my wrist, looking for the artery, and then the vein, to which Dr. Hayes would hook my cannula. I could feel some pressure and sometimes it felt a little wet, but it didn't really hurt. The procedure continued for about forty-five minutes. With my head and eyes covered, all I could do was pray and lie there and think. I envisioned my entire life up to that point and I suddenly wondered if I would ever see another Christmas.

It wasn't long before I heard Dr. Hayes say,

"Sutures, please." I knew this meant he was sewing me up and the operation would soon be over.

Then suddenly, without warning, I sensed something was wrong. I couldn't see, but I knew everyone suddenly quit working on me. There were footsteps, wheels rolling across the floor, and silence. It was an uncomfortable silence that hovered in every corner of the room.

Then I heard a nurse dialing the telephone.

"Please send a stretcher to surgery room number four. Also, page Doctor Battalia and ask him to call extension 436, please."

I tilted my head backward as far as I could and barely peeked out from beneath the sterile cloths that covered my eyes to see the wall clock behind me. Five minutes elapsed, then fifteen, and finally, after twenty minutes had passed, Dr. Battalia entered the surgery room.

"Mrs. Van Winkle, I'm Dr. Battalia. I'll be finishing up your surgery. Dr. Hayes suddenly came down with the flu and had to leave the room. We'll have you finished up here in a jiffy."

Oh, I thought, so that was the reason for all the silence. I settled down now because help was finally here.

It wasn't long before Dr. Battalia connected the two tubes leading from the artery and the vein, taped the tubes together properly, and then taped them down with a sterile dressing under and over them. Only a very slight portion of the cannula showed at the end of the dressing, just like Dr. Drake had told me, to enable me to keep a good watch on the color of the blood in the tubing.

"Now, your arm will be sore for a few days," said Dr. Battalia. "We're going to take you back to your room now and the nurses will keep your arm elevated and in ice packs all night. Tomorrow we will probably use the cannula and run you on a dialysis machine. I'll come by and check you in the morning. Try to get some rest tonight."

"Okay, thanks, Dr. Battalia," I replied.

As the orderly wheeled me away to my hospital room, I gave thanks for men like Dr. Hayes and Dr. Battalia who knew just what to do and when to do it.

Cannulas, such as had just been placed in my arm, usually last about six months to a year, and this operation turned out to be the first of about fourteen or fifteen similar operations to install new cannulas in my arms in years to come. I always hoped they would never run out of new cannula sites. The doctors assured me, however, that they could use my legs and put cannulas in them, too. Knowing this put my mind at rest.

The next day, Dr. Drake appeared in my room quite early in the morning.

"Good morning, Ann," he said. "We're going to help you to feel much better today. In a few minutes, the nurses will be hooking you up an artificial kidney dialysis machine. We will run you just six hours today and see how you do. We will send you home tomorrow and then have you and Jerry come into our outpatient dialysis unit the next day to start your training. Will it be any problem for Jerry to get off work for the day?"

"No," I replied. "I don't think so."

"Fine. Any questions?" he asked.

"Just one," I said. "My feet seem to be more numb today and I'm having trouble holding my fork and spoon and the coffee cup. I also tried to write something and I can hardly hold a pencil! Will this improve when I get on the machine?"

"I can't say for sure, Ann. We'll just have to wait and see," he responded. "I'll stop back a little later and see how you are doing."

"Okay," I replied. "Thank you."

As he was leaving the room, two nurses wheeled a big metal machine into my room.

The looks of this contraption reminded me of a portable, stainless steel dishwasher that might be found in the kitchen of a school cafeteria. It stood about three feet from the floor, was approximately four feet long, and two feet wide. It was equipped in front with more control switches, gauges and lights than the dashboard of an automobile. Underneath the controls a closed cabinet-like compartment contained many curious-looking hoses, motors, and pumps. On top of this whole thing sat three fiberglass-type boards with cellophane-type material—much like we use to wrap sandwiches in—between the three layers. Out of each end, between these layers, were two rubber tubes clamped off by some curious scissor-like tools that I later learned were called hemostats.

Above all of this equipment and dangling upside down on a metal post attached to the frame of the machine was a bottle filled with some clear liquid. Out of it, and clamped off with hemostats again, was another clear plastic tube.

"Good morning, Ann," said the nurse. "My name is Jeannie and this is Mary. How are you feeling today?"

"Oh, things could be better," I said.

"Maybe you will feel better in a few hours," Jeannie replied. "We're going to get this machine ready and dialyze you on it for a few hours today. Dr. Drake should be here in just a few minutes."

Jeannie prepared the machine by elevating the

three fiberglass-type boards, emptying a clear solution, attaching more plastic tubes which were about six feet long, and then clamping them off.

Mary, at the same time, was taking my blood pressure, counting my pulse, and preparing a syringe full of another clear solution which she inserted into the pump on my bed stand. As everyone was going about his specific duties, Dr. Drake walked in.

"Good morning again," he said. "It looks like things are going pretty well. This is a curious-looking machine, isn't it? It really isn't as frightening as it looks and I'm sure you will do just fine on it. The nurses will be hooking you up in just a few minutes. I'll let you go home after your six-hour run today and then I'd like to have you come back in two days and we'll start training Jerry how to do this."

Just then Jerry walked into the room. After greeting everyone, he bent over and kissed me good morning.

"How are you doing?" he asked.

"Pretty good," I answered. "How about you and Jeff ?"

"Oh, we're fine. I fixed his breakfast this morning and I think he had more mush on the outside than on the inside," he said. Everyone laughed, breaking the tension.

"Okay, now we're going to unwrap your arm and get going here," said Jeannie.

As Dr. Drake looked on, she carefully unwrapped my arm bandage, released the tapes holding my new cannula securely together, and, with some small wire clamps, clamped off the tubing protruding from my arm. When this was done, she removed a plastic connector that joined the two tubes, connected one six-foot-long line from the machine, and released the clamp on the arterial side of my cannula. I watched the blood flow out of my arm and into the machine.

My heart began to pound a little harder and faster as I saw my blood escaping my body through a clear plastic tube. Panic wanted to set in but I didn't allow it. By this time, the pumps and engines in the closed cabinet on the bottom of the machine were clicking a happy rhythmical tune. As the blood flowed into the machine, it pushed the clear solution in the cellophane paper out the other end and into a waiting pan. When all the clear solution was pushed out and blood was coming out the other end of the dialyzer, Jeannie hooked the second six-foot-long plastic tube to the other side of my cannula which was called the venous side. It would carry the freshly cleansed blood back into my system to be recirculated. By the end of the six hours, my own heart would have pumped the blood from my body, through the machine to be cleansed and back into my body several times. However, during this six-hour run, there would never be more than a cup of blood outside my body at any given time. As a cup was going out, a cup was coming back in. This circular procedure would continue uninterrupted.

"Looks good!" said Dr. Drake. "How do you feel?"

"Okay," I said, "just a little nervous."

"Did it hurt?" he inquired.

"No," I responded. "I didn't really feel a thing."

Turning to Jerry who was intently watching the whole procedure, Dr. Drake remarked, "We'll start teaching you how to do this procedure day after tomorrow. Think you'll be ready?"

"Sure. We'll give it a good try," Jerry replied.

"Now, Ann," said Dr. Drake, "don't be afraid to

move a little. If you should kink your blood line, or do anything else wrong, a buzzer alarm will go off and this light will flash. You may sleep if you want, eat lunch, watch TV, read, talk on the phone, or do about any other quiet thing to pass the time. I'm going to take your blood pressure now and I'll stop in during the day to check on you. The nurses know where to find me if they need me."

I lay back to wait, watch, and listen for the next six hours. Jerry stayed with me and settled down to read the morning paper. I couldn't concentrate on anything except this machine which was swallowing my blood.

I was still watching it when lunch arrived several hours later. It was a welcome relief to eat, one of my favorite pastimes.

All kidney patients had to be on a special diet. Because my kidneys were not eliminating the fluid taken into my body, my total fluid was limited to one-half cup water, coffee, jello, ice cream, pudding, or pop per day. Anything that turned into liquid would be held in the body until the next dialysis time.

I was also limited to a low-sodium diet. No more salt shaker. I would learn in days and weeks ahead just how many more "no-no's" there were in this department. Carbohydrates were not limited, but protein was. This, too, I would learn more about, I was told.

So, on the tray before me was a typical lunch:

Green lettuce salad with a
wedge of tomato

Salt-free salad dressing

Buttered macaroni

Fresh cauliflower

Small ground beef patty

Bread & salt-free butter

Sponge cake

Three pieces of hard candy

While Jerry excused himself to go to the cafeteria, I nibbled at my food. Eating with tubes attached to my "eating arm" was rather awkward. I found the alarm on the machine would go off when I moved, so I ended up eating with my left hand. This definitely left something to be desired.

With lunch half eaten, I settled down again to watch the clock and the machine—two-and-one-half hours until my "coming off" time.

And then, I heard someone walking down the hallway toward my room. Uneven, unsure footsteps echoed through the wooden-floored, dark corridor.

As the sound neared my doorway, I sat up in bed, encumbered by the tubes that protruded from my arm leading to my new kidney machine.

A woman stood in the doorway and looked in my direction. Her lean body, encased with yellowish, drawn skin, leaned against the door jamb. She was wearing faded blue jeans and a yellow plaid shirt that hung limp and lifeless. Her long brownish hair was pulled back and tightly bound high on her head. It swung like a pendulum as her head bobbed forward. She wore a chain around her long slender neck and on the end hung her cannula clamps. I knew from this evidence that we were both kidney patients.

However, as her sunken eyes stared in my direction, I knew we were different. She was blind. It was another result of kidney failure and my mind dashed through thoughts of this possibly happening to me. At her side she carried a cane.

"Hello," she said in a weak, tired voice. "My name is Pat. Just thought I'd come in and say hi. I just got off the machine down the hall. Guess we're walking down the same road these days. It's hell as far as I'm concerned. I don't like being blind and not being able to do things on my own. It's all because of this blasted machine. I don't know about you, but some days, I think I'd rather be dead!"

"That's too bad," I answered. "I know it's not easy, but right now I think it's worth it. My little boy is almost two years old and life is worth living. I'm going to fight it out. It's tough, but I think it's worth it."

She turned and left without a word. I knew we were on different roads then. She would have to walk hers and I would walk mine as best I could.

After this sad encounter, the time went by fairly quickly. Jerry came back and we talked, Dr. Drake stopped by again, nurses were in and out, and my mom stopped by to see how I was doing. Finally three o'clock came and I was unhooked and set free once again. A reverse procedure from putting me on the machine was performed for the unhooking. Jeannie did only one thing differently. Instead of the three fiberglass boards lying horizontally for the "take off," she tilted them so they were in a slanting, vertical position. This, she explained, was so all of the blood would run back into me by gravitational force. She then clamped off my arterial cannula, unhooked the plastic line from the machine and held it up in the air.

The blood immediately started through the dialyzer and back into my arm through the venous cannula. As soon as all the blood was back into my body, the line was clamped off, unhooked and my cannula was put back together again. The blood then continued to circulate through my body and this tube lay, covered by sterile dressings, on the outside of my arm. First, however, before my arm was wrapped, the cannula sites, or the holes where the tubing was hooked into my arm, had to be cleaned with alcohol and a medicated creme put around each site. It was very necessary to keep this area meticulously clean, because it was a good breeding ground for infection.

When the nurse had finished, she said, "Now, I want to take your blood pressure and then you can get up and weigh on these scales. From now on, we will record your weight before and after dialysis to see how much fluid the machine draws from you."

This machine was not only taking the wastes out of my blood, but was drawing off extra fluid, and putting good, needed chemicals back into my system. What a miracle! And to think that I was one of the first ten in Oregon to pioneer this field of home dialysis. At the time, I didn't realize what a privilege this was.

Others who pioneered with me were not as fortunate. Jim died from eating an overabundance of foods high in potassium. Cindy's husband left her to fend for herself. (He couldn't stand the pressure of having his wife on a machine.)

Then there was Liz. She was a beautiful, shapely blond who couldn't tolerate thoughts of losing her beauty to kidney disease and its pitfalls. Liz threatened suicide and finally succeeded in ripping out her cannulas and bleeding to death.

I dismissed these ugly thoughts from my mind very quickly.

I climbed out of bed, literally, for our beds were up on ten-inch blocks. The philosophy was that it was easier on the heart if it pumped blood downhill to the machine instead of uphill.

I weighed, got dressed, and Jerry and I bid the nurses good-bye, saying we would report to the outpatient dialysis clinic in two days. For the moment, we felt free as we left the hospital. I felt very weak after my dialysis, but I pushed myself to keep going. At home was my little Jeff, and we looked forward to our family being together once again.

When we arrived home, Jeff was full of jibber jabber about the events that had happened during the day. He had been with a babysitter. His second birthday was coming up very soon and I questioned him about what he wanted.

"Ball," he would say. "Play ball, Mommy."

He would roll the ball to me as I sat completely washed out on the davenport. He loved balls and I believe if he had owned a dozen balls, he would have wanted one more.

In my mind, I was making plans to have a second birthday party for him out on the back deck. I thought this might be the only birthday party I would be able to give him.

My future, at this point, was very uncertain. I knew it, the doctors knew it, and Jerry did too but didn't want to admit it. This first year of dialysis, which was just beginning, was to be the hardest.

After we ate dinner, which had been provided by some wonderful church ladies, Jerry did the dinner dishes while Jeff and I sat down to read his favorite story, *Peter Rabbit*. The time at home flew by all too quickly and before we knew it, the second morning was here. The time had come to go back to the hospital for dialysis treatment and training. This time Jerry would be setting up the machine and putting me on it. I had a monumental amount of faith in his ability to do everything correctly.

In fact, I had great faith in my husband to do almost anything. We had known each other since high school. I knew we would some day be married. But first, came the growing up experiences: football games, dances, parties, good friends, and then separation. Jerry graduated one year ahead of me and enlisted in the U.S. Marine Corps. How handsome and manly he looked in his dress blue uniform!

We were apart for two years before he returned home. There was college ahead for me and a career in piano.

When Jerry returned from the service, he was stronger and more grown up. I admired his ability to endure tough situations and his sensibility. When he asked me to marry him, my answer was yes. Little did we know how true our marriage vows would be. Yes, we would love and honor each other through sickness and in health as long as we both should live.

It was not an easy test. Now, I was really trusting him with my life. He would run me on the machine for eight hours today.

I awoke early in expectation of the day ahead. Dr. Drake had ordered the eight hours of dialysis treatment for me today, preparing me for eventual ten-hour runs on the machine. I would run three times a week, which would make a weekly total of thirty dialysis hours.

As I lay there in deep thought, I could hear stirring from the other room. Jeff was waking up. I threw back the covers, swung my weakened feet and legs over the edge of the mattress and onto the floor, stood up, and collapsed in a heap onto the hardwood floor beneath me!

"What happened?" Jerry asked as he awoke, startled by the noise. "Are you all right, honey?"

"I'm okay," I replied. "I guess I just slipped."

This was the beginning of many more unwarned collapses onto the floor. My weakened muscles were giving out quickly from unresponsive nerves that had been damaged from the poisons in my system. These uremic poisons had built up from the result of not being dialyzed properly nor soon enough at the beginning of my illness.

I picked myself up from the floor, tended to Jeff, fixed a little breakfast, and we were on our way. We dropped Jeff at the babysitter's house and drove on into Portland for our dialysis training. I felt a little puffy and itchy, like I had too many poisons in my system and something needed to be rectified.

We arrived at the center, found our way through a basement door of the old nurses' dorm at Good Samaritan Hospital to an ancient elevator that was to carry us to the second floor—the outpatient dialysis center. We pushed the button to the elevator and nothing happened.

"Just push it again," said a passing custodian. "That thing is really tempermental but it works—once it gets here!"

We followed his instructions and sure enough, the rickety, old elevator arrived. We opened one door to find a heavy wire-like door that had to pushed back like an accordian to let us in. It was almost more than a one-person operation. The elevator, once we were inside, carried us to the second floor. We used the reverse procedure on the doors and got off—both saying a silent prayer of thanks that we had safely arrived and had not been stuck between floors somewhere.

Jeannie greeted us.

"Good morning," she said. "All ready for a big day?"

"Guess so," we replied almost in unison.

"Come right down this hall," she motioned. "Ann, you weigh on these balance scales right here and record your weight on this chart. As soon as you've finished with that, you may crawl into bed and I'll show you the correct procedure to clean your cannula sites."

Turning to Jerry she introduced him to Bob, our machine technician. He would show Jerry all the mechanical features of the machine, help him to get it ready, and hook me up to it. Everything was done just like clockwork, and each step in preparing the machine was checked off against a list of procedures.

Finally, it was time to hook up my "life lines" again and Jerry was elected to do this. He followed instructions, and in no time at all, I was "off and running," as the old dialysis phrase goes.

"After your run today," Jeannie said, "the machine has to be cleaned. Bob will show you this procedure, Jerry, on a dirty dialyzer that we already have. You can be doing this while Ann is running her eight hours."

"Fine," he replied. "Any time."

As the days and weeks continued, we would learn many new "do's" and "don'ts" for dialysis patients. We were willing to learn all we could because we realized that to get the maximum benefit from the machine and for my life, we had to learn such things as: how to declot a cannula when the blood flow had stopped due to a blood clot getting caught in an artery or vein; how to handle blood pressure drops by administering saline; how to give blood transfusions through the machine; what to do in case my cannula separated accidentally; and lastly, what I could and could not eat. Reluctance to learn the new rules was written all over my appetite!

Bless me with life so that I can continue to obey you.

—Psalm 119:17

Chapter 5

Bye-Bye
Salt Shaker

"The first important rule about your new diet," said Jeannie, "is virtually no salt at all. Nearly all food contains some natural sodium or salt, but you need to learn which foods have additional salt and learn to stay completely away from them. You can say bye-bye to the salt shaker on the table because it is a real 'no-no'! Don't worry, though, you'll become used to eating unsalted food. Most patients tell me they don't mind it after a while."

Just then Gloria entered the room. She had been on dialysis for three years, I learned, and was helping out at the center. In fact, I also found out that this was the Gloria I had sent money to several years ago when I heard the radio announcer's plea for financial aid on her behalf. Now, here she was helping me!

"It's true what Jeannie says," Gloria replied. "I'm pretty used to a salt-free diet now. Of course, I cheat once in a while, but not too much! I'm not the only one, though. The other night Jody washed the salt off of some potato chips and stood there popping them into her mouth before they got too soggy! Then, she went on dialysis." We all laughed.

"Why is it that we can't have salt?" I questioned.

"Well, in the first place," Jeannie said, "you know that if you eat something like potato chips that are salted, you get very thirsty, and, already you know that your fluid is limited to four ounces or one-half cup a day. If you eat salty food, you would naturally want to drink more. If you drank more fluids, they would be held inside the body until your next dialysis. You would get puffy and itch where the fluid collected in your tissues. Also, your blood pressure would go up and this would make you a high risk for a heart attack or stroke."

"Oh, that doesn't sound too good," I responded.

"No!" Jeannie replied, "So you see, it is very important that you follow the 'no salt' rule. Now, let's look at this list of things you need to avoid."

At the top of the list was salt. No more salted eggs in the morning, no salt in the pan before cooking the frozen vegetables, no salt in or on mashed potatoes and gravy and no salting the fried chicken! The "no salt" rule, I learned, applied to anything with the word sodium on the label—even monosodium glutamate! Also, the rule applied to such luxuries as celery salt, onion salt, and any other condiment such as Worchestershire sauce, barbecue sauce, salad seasonings, ketchup, mayonnaise, mustard and the like.

Next, I learned the rule applied to almost anything and everything in a can, except canned fruit, fruit juice, and canned goods that were labeled "salt free."

My goodness, I thought, this is practically everything in the grocery store. No more canned green beans, soups, corn, beets, tomato sauce, baked beans, tuna fish, chili, canned lunch meats, shrimp, crab, peanut butter, or anything that was commercially canned.

After discovering these sad facts, we turned to look at the list of frozen "no-no's." At the top of the list was written: "No commercially prepared frozen convenience foods." This included no frozen pizzas, chinese food, frozen vegetables, pies, frozen convenience meats, etc.

No pizza! I thought. This is getting to be too much. I loved pizza!

Next, we covered the meat case. Here all the lunch meats such as bologna, salami, chipped beef, cheese (including cottage cheese), hot dogs, all sausages—unless labeled salt free—ham and bacon were absolutely excluded from my new diet.

I was beginning to wonder what I could eat when we began to cover the bakery items prohibited. For the time being, I was allowed to have bread and butter that contained salt, but this did not last long. Shortly I would be learning to eat even these two items "salt free."

In essence, "salt free" practically became my middle name. After all of the "don'ts" however, we got to the "do" list. Finally, something that I could eat!

The main contents of this list included all fresh fruits, fresh vegetables and fresh, uncured or unsalted meats. This did include a nice variety of food.

At the vegetable counter, I could buy and eat all the green leafy vegetables such as spinach, chard, beet greens, lettuce (which I love), and cabbage. Then there were fresh carrots, beets, broccoli, celery, zucchini, cucumbers, mushrooms, brussel sprouts, parsnips, and more. This didn't seem so bad after all. I always did like vegetables, so maybe I would survive.

Although fresh fruits were limited because of the potassium content in them, I was allowed to eat most of them. Apples, cantaloupe, fresh pineapple, and grapes (green ones, of course) were my favorites in this category. Oranges, bananas and tomatoes were high in potassium so I didn't get very many of these.

At the meat counter, choices had to be made very carefully. Ham, bacon, sausage and sodium-filled seafood were not allowed. Much of the other meat was allowed, but only in small quantitites because it is harder to assimilate by the body. When eating too much protein, the BUN (blood urea nitrogen) rises and makes you itch all over, especially up and down the spine. My choices, therefore, were usually hamburger, chicken, turkey, or roast.

Jeannie told me that I must watch out for foods loaded with potassium. When the kidneys don't work, the potassium builds up in the blood. "If it gets too high, or too low," she explained, "the heart could stop beating."

Patients were known to have killed themselves by eating too much potassium.

Bananas, oranges, tomatoes, raisins, orange juice, baked potatoes, and french fries are very high in potassium and should be eaten very sparingly. Once, I was told, a patient ate half a box of raisins all at one time and died.

One of the complexities about my new diet was being invited out to dinner and maybe not being able to eat what the hostess had prepared. Usually, before the night of the dinner, the hostess would ask what I could or couldn't eat. I would tell her to just fix a plain roast with fresh vegetables and not to salt anything. This usually worked out well.

Restaurants were a different situation. Some were very accommodating and others were not. I will never forget a restaurant we went to at the beach. Being hungry for prawns, we inquired if, by any chance, the cook had a salt-free batter they could be dipped in and then fried. The waitress said that she would ask. In a flash, out came the cook to our table. He wanted to know just what I could and couldn't have. Upon discovering my need for a salt-free batter for the prawns, he said to just leave it all up to him; he'd be glad to fix them like I wanted them—and he did. They were delicious and a rare treat! Usually in a restaurant, I would order a small, unsalted steak and a green salad. Nearly everything else was salted or too high in potassium. How I missed the baked potatoes and french fries!

One item that I could eat a lot of was hard candy. The sugar, in some way, helped me a lot. It decreased breakdown of muscle tissue and added calories to my diet. Wouldn't you know it—I never was much for candy, but I did my best to eat my daily allotment.

Cooking all of these special dietary requirements at home did not become much of a problem. Jerry and Jeff were usually content to eat whatever I was eating. They used the salt shaker at the table instead of having the food salted during cooking time. Every now and again, they would get to eat things I couldn't have, like pizza and spaghetti. Even these, however, can be homemade without using salt. Fresh tomatoes can be made into a sauce and salt-free cheese is available at the grocery store.

As we all became used to this new way of life, many adjustments had to be made. Times were not fun and easy at all, but our faith, hope and love was keeping us together. We prayed that somewhere, on down the road, tomorrow perhaps, our shining star would glow and God would grant us better days.

Chapter 6

Wheelchair Days

*I*t was a sunny afternoon in late October, 1967, only two months after going on the kidney machine, when it happened again. I had been sitting on the living room couch waiting for my friend Marge to arrive. Jeff, our busy, energetic two-year-old, was doing something very quietly in the kitchen. Too quietly, as a matter of fact! As I got up and walked toward the kitchen to see what he was up to, my legs gave out, and I landed in a heap in the middle of the living room floor.

This uncontrollable falling had happened three or four times now. The last time was just the week before. Now, as I sat in the middle of the living room floor, these previous scenes rushed through my mind. Jerry was not home to pick me up, and Jeff was certainly too little.

Trying not to upset Jeff, who came running into the living room when he heard the big noise that my crumpled body made when it hit the floor, I blinked back the tears.

"Mommy fell down and went boom!" I tried to say jokingly.

He stood there with a real puzzled look on his face until I started to laugh. Then he laughed too, knowing everything was going to be all right.

Deciding not to trust my weakened legs to lift me off the floor, I chose to crawl over to the davenport where I could boost myself up with my arms and sit down.

Just as this feat was accomplished, Marge drove in the driveway. Looking through the big picture window behind the couch, I could see that she had brought Blake, her son, with her. Jeff and Blake were only about one year apart in age and they would have a real good time together. This would give the two of us some time to really talk. We had a lot of catching up to do. Marge and I had met during college days at Portland State. We both worked in the library and developed a close friendship which we didn't realize would last for many years to come.

Marge was the kind of person who sometimes is put on a pedestal. To me, she was a superstar Christian and had everything all together. She also knew just what to say and what to do at the right moment. It looked like today I would be pouring my heart out to her. We hadn't seen each other since I had been on the kidney machine so there was much to talk about. My ending up in the middle of the floor was the most recent situation that I needed to talk about, so I could see she was in for it. God had sent her to our house today, just at the right moment.

"Jeff," I said, "can you be a big boy and open the door for Marge and Blake?"

Just then the doorbell rang and Jeff ran to the door to carry out my wishes. As he flung open the door,

Marge stood there with arms full of all kinds of good, fresh vegetables from her garden.

"Hi, hon," she began. "My, what a big boy you are, Jeff. Thank you for opening the door. Do you remember Blake?"

The two boys stood looking at each other for a minute, but very soon Jeff was off to show Blake his new tricycle and the new puppies that were out in the back yard. We would not see them for a while, since nothing goes together better than two boys and a litter of playful puppies!

"Sorry I'm not getting up off the couch," I said as my eyes began to well with tears. "I'll have to tell you all about it."

"Okay. Just a jiff," she replied. "Let me go put these vegetables down in the kitchen. I hope you can use them."

"Use them? You bet I can!" I replied. "My middle name is 'vegetable' since I've been on this new diet. We will love them. Thank you so much!"

After she laid the yummy carrots, zucchini, beets and beet greens on the kitchen counter, she came in and took a place beside me on the couch. We exchanged hugs and it was all I could do to keep from bursting into tears.

"It's so good to see you," I commented sincerely. "You must have known that I needed you today."

"I'm so glad that I came," she replied. "You're looking pretty good. Now, tell me, how is everything going?"

"First off," I said, "you know I've been on the kidney machine almost two months now. In fact, we are almost through with our training at Good Samaritan Hospital and we'll be moving the kidney machine home in about a week. It's kind of scary. Jerry has been super to learn the setting up and

cleaning of the machine, Marge. I just couldn't ask
for anyone better, but I know it's hard on him. I'm
really lucky that he is sticking by me. There is
another girl about my age on the machine at Good
Samaritan and her husband wants a divorce! Imagine that on top of being so sick. I can't even imagine
how horrible it would be to have to go through a
divorce too! It just doesn't seem fair. I don't think
that I could handle that. But one never knows. They
say that being on a kidney machine is really hard
on families and marriages. So, I just hope and pray
that Jerry and I can hang in there!"

"Don't worry," Marge replied. "He'll stick by
you!"

"I hope so," I answered. "I sure do love him."

"He loves you, too," Marge answered. "He loves
you very much."

"Also," I continued, "now that the kidney
machine and my new cannula are becoming a way
of life, there's another problem."

"What's that?" she asked.

"My legs have been giving out on me. Just before you came today, I got up to see what Jeff was
doing and I ended up in a heap on the floor. My legs
just gave out and down I went! I never know when
it's going to happen!"

"Have you talked to the doctor about this?" she
asked.

"The doctor knows I've been having trouble with
my feet and hands. I've been dropping things too,
and playing the piano is just not for me anymore.
It is really discouraging because I don't have enough
strength to even push down a key on the piano. The
doctor doesn't really know what the answer is. They
have been dialyzing me twelve hours at a time, three
times per week in hope of restoring the nerves that

have been damaged from uremic poisoning. Only time will tell if this is doing any good.

"Very soon now, Dr. Drake is turning me over to our family doctor, Dr. McMahon, who will be following my case. The load gets too heavy for one doctor to handle all the kidney patients, so they try to get other doctors to help out and then just report any problems to them. In fact, Jerry and I have an appointment with Dr. McMahon tomorrow. I'll tell him about my legs and see what he says. This wondering and waiting is really bugging me. I'm getting afraid to walk anymore, for fear of falling down!"

"You're really going through a lot of trying times right now," Marge said. "Just remember what the Bible says in Romans 8:28: *All things work together for good for those who love the Lord and are called according to His purpose.*

"I know that you love the Lord," Marge said. "We don't know why all of these things are happening to you right now, but just remember that God still loves you and has a wonderful plan for you and Jerry. Just trust Him and hang in there!"

"I know that I am really lucky," I replied to her. "You know a few years ago, I wouldn't have had the opportunity to go on a kidney machine. They didn't even have this program in Oregon. In fact, do you know that the Kidney Association of Oregon was set up only three months before I went on the machine? That just blows my mind when I think about it. If that organization weren't here, no way could Jerry and I afford all of this. So, you see, I really am a lucky person."

"I know you are," Marge said. "Tell me, what does the Kidney Association do?"

"They buy the kidney machines for home dialysis

and then loan the machines out to the patients. They help financially with supplies needed to run the machine and with my medicine, too.

"The association is most willing to help in any way they can. They have a big job, though. Already, in 1967, they have ten of us on home kidney machines. That really gets to be expensive. They say it costs as high as $750 to $780 per month considering hospitalization, medicine, equipment, and supplies. Before KAO [Kidney Association of Oregon] was in existence, however, candidates for treatment had to guarantee $30,000 in advance and $10,000 per year after the first three years before they could even accept them for any kind of dialysis.

"So, you see, I really am lucky. We just couldn't come up with all that money!"

"God works in mysterious ways and we never know when or how," Marge replied. "We just have to have faith. Hey, how about a cup of tea?"

"No tea for me," I said. "I can't have the liquid, but please fix one for yourself. I'll tell you what. I just crave ice cubes, so you could bring me an ice cube in a little glass, please. One ice cube lasts longer than drinking that much water, I've discovered. I don't know why I crave the ice, but I sure do!"

"Okay," she answered. "One ice cube coming up. I'll check on the boys, too. They're being unusually good. Those puppies must be just the ticket!"

As the afternoon flew, Marge and I were still in deep devoted conversation when Jerry arrived home.

"Anybody home?" he called as he entered the house from our garage.

"Just us," I answered.

"Oh, hi, Marge, I wondered whose car was out front. Haven't seen you for a long time. How's Ben?" Jerry asked.

"Just fine," she replied. "He keeps real busy with his seventh-grade class. He loves it."

"Good, tell him hello for me," said Jerry.

"I'll do that," she assured him. "And speaking of Ben, I think Blake and I had better get home and fix Daddy's dinner. We could be replaced if we're not careful!"

"I'm so happy you came over," I said. "Don't wait too long to come back. You're always a bright and shining star in my day and a real inspiration."

"No, lady—you're the inspiration. Now, you just keep your chin up and everything is going to work out. Let me know what the doctor tells you."

As I related the happenings of the day to Jerry, he fixed the dinner and put all of the fresh vegetables away. I could tell that something was on his mind, so I decided to ask him.

"Is something bothering you?" I questioned.

"Oh, not really."

"Are you sure?" I asked again.

"Oh, it's just that I've been thinking about you being here alone with Jeff during the day. I think maybe we should get someone to come in and help you. You're not really up to cleaning house and everything. Maybe they could do a little of that too," he said.

"Do you have anyone in mind?"

"No," he answered, "but maybe there is some agency that does this kind of work. We'll ask the doctor tomorrow."

"All right," I replied. "I love you, honey."

I leaned over to give him a kiss and we held each other very tightly for a few moments.

The evening passed all too quickly. Jeff was getting to be such a big help. He helped clear the table and then got his pajamas for me to put on him. Then it was time for the *Peter Rabbit* story, which we had heard for over the hundredth time. He loved to hear all about how Peter went to Mr. McGregor's garden and got into trouble. He would sit spellbound throughout the entire story that he practically knew by memory. At the proper time, he would always insert the words that he loved the most.

"Okay, time for bed," I said. "Come on, Daddy, let's go say prayers and do 'aaah.' "

The three of us would often stand hugging in the middle of Jeff's blue and white bedroom saying "aaaaaaaah" all together on one breath and then we would tuck Jeff into bed and say our prayers. This time together was really the highlight of everyone's day. We were a close family and this seemed to make all the new experiences we were going through a little easier to bear.

Six-thirty in the morning came all too early. It was going to be a full day. Jeff was still asleep in his crib, so I decided to get an early start and take my shower first. It was taking me longer to get ready these days, because I needed to hang onto walls, doorways, dressers, and anything else at hand as I walked. I was terrified of falling down again!

We made it through the morning without a crisis, delivered Jeff to the babysitter, and reported into the outpatient dialysis clinic at Good Samaritan by eight in the morning.

Jerry did a fantastic job of putting me on the machine that morning. The nurses said we were about ready to start home dialysis. The whole thought of being on our own was frightening, but we knew the time was approaching rapidly. Only a few more days and they would deliver a kidney machine to our house. It would be up to us then. The nurses would be on call all night long if we needed them. However, as reassuring as this was meant to be, there was still that fear of being at home, on our own.

The day went by as usual, except I had a little shorter run. Our appointment with Dr. McMahon was at four-thirty. In addition, this was my day for a blood transfusion. When the kidneys are not functioning properly, they do not make a hormone that stimulates the bone marrow to make red blood cells. As a result, my red blood count ran fairly low—a number of about eighteen to twenty—and if it got down around fifteen, I had to have a blood transfusion. This blood was given to me about every two weeks. Jerry would just plug the bag of red cells into a drip bulb on the kidney blood lines, the plastic tubes that hooked onto my cannula. The packed red cells always made me feel better, so I would be in pretty good shape today when we went to see Dr. McMahon.

The time came for me to come off the machine. Again, Jerry did a fine job of taking me off and we were on our way. Jerry would clean and sterilize the dialyzer on my next run. He always kept one clean, sterile dialyzer ahead and this was an advantage of running in the hospital. We would have only one dialyzer at home, so it would have to be cleaned and sterilized immediately after each run.

We arrived at Dr. McMahon's office on time and I eagerly took a seat in the attractively decorated waiting room. Coming off the machine usually made me feel very tired. It was not unusual for me to push myself to keep going. A chair looked very inviting right now, so I plunged right into the middle of it. Not for long, however, for momentarily the doctor's nurse called us into his office. He got up as we entered and seemed very glad to see us. He truly was our family doctor, for in his younger years he had taken care of my grandparents. He knew our whole family and it seemed like he really cared about us personally.

"So nice to see you," he said as he greeted us. "I've just been talking to Dr. Drake about you. He says you're doing fine on this new kidney machine adventure. How do you feel about the whole situation?"

"We're becoming accustomed to it," Jerry answered. "It isn't easy, but then no one said it would be. We're getting along pretty well."

"Jerry has really helped by learning to run me on the machine," I commented. "I couldn't do it without him."

"I'm glad to hear that," said Dr. McMahon. "It takes two strong people to go through such a trial and not give up the ship. I admire you both. Now, you're mainly here to talk about my being your follow-up doctor, I understand. I'll be very frank with you, as I was with Dr. Drake. I have had no experience with kidney machines, but I am willing to learn. I've been reading up on dialysis, but have never actually run a machine. I told Dr. Drake we'd be glad to test your blood after your runs and to follow any complications that might arise, but, as far as running the machine, you probably know more about that than I do."

"You really don't need to know how to run the machine," I said. "We, or, I should say, Jerry is learning to do that and we just need your help for other things that may arise. For instance, I do have a problem right now. I'm afraid to walk anymore. My legs have been giving out on me and I fall in a heap on the floor at almost any unpredictable time."

"How long has this been going on?" he asked.

"Well, it's been pretty bad for the last two weeks. My toes and fingertips have been getting numb for about three months now, but this falling really has me worried," I said.

"May I test your reflexes?"

"Sure."

He tapped below my knee and behind my heel with a small hammer and stuck my feet and legs in several places with a pin.

"Can you feel that?" he would ask.

"No, not there," I replied.

"There?" he questioned as he moved it up toward the knee.

"A little better," I replied.

After much testing, he turned to me.

"Ann, for your own safety, my recommendation is that you start getting around in a wheelchair. Jerry is not going to be with you all of the time to see that you don't fall. It would not be good for you to fall and break your leg, or a hip, or an arm at this point. Do you understand my concern?"

My mind was already off in a world of its own.

A wheelchair! The thought raced through my mind. I don't want to be a cripple. A wheelchair! A cripple! A wheelchair! It was all I could think about for the moment. I could hear Jerry and the doctor talking in the background, but it was as if I wasn't even there. Finally, I came back to my senses.

"What is the problem with her legs?" I heard Jerry ask.

I tried to blink back the tears as I leaned forward to listen intently.

"I'm quite sure Ann has what is medically known as peripheral neuropathy. It has been brought on by having too many toxins in the bloodstream. These poisons have destroyed the nerves that in turn signal the muscles to function. When the muscles don't get the signal, they begin to atrophy or waste away from lack of use. Now, we are trying to reverse this situation, as I am sure Dr. Drake has told you. Good dialysis is a must and we also need to start you on a therapy program. For this reason, I would like to admit you to Emanuel Hospital for a time. They have an excellent therapy rehabilitation program."

"Go to the hospital?" I asked. "When?"

"As soon as you can," he answered. "We'll be able to watch you closer in there and get you started on the therapy. Sunday afternoon would be a good time to be admitted. Do you think you could make the necessary arrangements and be there then?"

"The only problem is that I'm supposed to run on the kidney machine on Sunday," I answered.

"That's all right. We'll make arrangements to have your machine moved right into your room at Emanuel. We'll dialyze you there Sunday night," he said.

"All right, we'll be there."

"Any other questions?" he inquired.

"Well," Jerry interjected, "we were wondering if there is some agency where we could get some domestic help, but it doesn't look like we'll need it for a little while. I will just arrange to take Jeff to the babysitter and think about getting some help for Ann when she gets home from the hospital."

"Yes, there is a good agency and I'll get the name for you. You'll probably need someone if you don't have some family member that could help out. We'll talk about it later," he reassured us.

As we stood to leave, I hung onto Jerry very

tightly in case my legs might give out again.

We left the office almost in a state of shock. Where were we going to get the wheelchair and how was this going to fit into our already unpredictable future? It was just one more problem. Somehow, some way, we would manage it.

Jerry rented a wheelchair for me. Jeff thought it was great fun to sit on my lap and have Daddy push us around. It wasn't so much fun for me—mostly because I had always been so independent, and now, here I was, dependent upon other people to help me get around. My frustration began showing by the time the weekend was over. It was almost a relief to check into the hospital that Sunday evening. The feeling that help was on the way once more cheered me. Surely, I would be out of this wheelchair in no time. After all, getting used to being on a kidney machine was enough for now. This wheelchair, although apparently needed, was very much *unwanted*.

Listen to this wise advice; follow it closely, for it will do you good, and you can pass it on to others: **Trust in the Lord.**

—Proverbs 22:17

Chapter 7

Stepping Stones

*B*ecoming accustomed to a hospital routine again had its good and bad points. It was a happy, contented feeling to be where help was readily available if and when needed. However, to leave Jeff and Jerry again for another stay among sterility, a private two-bed room (no one occupied the second bed), needles, syringes, unknown faces, hospital schedules, shifts, medicine, and institutional food, left much to be desired. In spite of the above, settling in took place and lasted four to six weeks. This time span was unknown at first—not even the doctors knew the future for me. It depended upon my progress, or lack of progress, as the case might be. The doctors were at a loss to know if the paralysis taking over my feet and fingers would continue on to paralyze my whole body, or if it could be arrested—maybe leaving me in a wheelchair for the rest of my life. I did not know until months later that the doctors were predicting that I would probably never walk again. God, in His wisdom, kept me from hearing these words, and for that, I am so grateful.

One of the first encounters I had after meeting the nurses and having Jerry run me on the kidney machine in my new room was with the therapy department. It seemed as though my legs and hands were getting weaker by the hour. The therapist would come to the room and give me exercises to do on my bed. He would actually lift my legs and maneuver my feet to make them work. Other hours in the day were spent by myself, doing the exercises that he taught me. When I wasn't exercising, I was rolling along the long hallway in my wheelchair or running on the kidney machine. There was a mirror in the long hallway and as I rolled past it, the reflection of a person in a wheelchair stared back at me. For a fleeting second, the reflection was someone else, not me!

Dialysis, blood tests, and therapy followed each other like leaders in a game. "Will this nightmare ever go away?" I wondered. "Am I getting worse instead of better?" It was very difficult now to feed myself, and writing was a chore not easily done at all.

I escaped from the wheelchair into my bed for many hours during the day. My mind would wander and wonder. Time was heavy on my hands and my heart was sad. As I lay there looking at the round, black and white institutional clock, a poem rang out from within me and flowed onto a handy piece of scrap paper:

TIME

Time is the steady tick of the clock,
Days passing by, years rolling on.

Time is a flower seed, beginning to bloom,
An old man, God's plan fulfilled.

Time is a baby's first breath,
His first smile, his first step.

Time is suffering and pain,
Trials and tribulations.

Time is remembering
happiness, laughter, gay times.

Time is days past and days to come.
It is for a moment and for a lifetime.

Time is eternal.
Time is for rejoicing.

Time is for regretting.
Time is a gift.

Time is from God.

Breaking my train of thought, the therapist entered.

"Good afternoon, Ann," he said. "You look like you're deep in thought."

"Oh, it's nothing. I just wrote a poem and it was on my mind," I replied.

"May I hear it?" he asked.

"Sure," I replied and read it to him.

"That's very good. May I give it to the nurse to hang on the bulletin board in the hall?"

"I guess so," I answered.

"Good," he said. "Now, I've come to tell you two things. First, Dr. Drake will be coming over Wednesday, and he would like to see you in the conference room.

"Second, on Wednesday, we are going to start working with you in the therapy room. You will be able to work on more equipment down there and strengthen your legs and hands. You'll also be going to the vocational rehabilitation center that is downstairs by our department. You'll have a chance to make art projects, type and do other activities to strengthen your hands. Do you have any questions?"

"I don't think so," I answered. "It sounds like Wednesday will be a full day and something that will be fun. Tomorrow I run on the machine again, but I'll look forward to Wednesday."

As he left the room, I picked up my Bible to do some reading. Why is it that God has to literally hit us over the head before we turn to Him. My earthly lessons were being learned the way God had planned. Daily I was learning step by step to turn my life over to Him. The lessons were not easy and many tests were yet to be revealed.

The afternoon faded with a glowing silver gray sky and the rain danced against my open window. Getting out of bed and into my wheelchair to go close it was too much of a chore, and I decided to leave it open when in walked Jerry and Jeff. The pleasure and joy that my two "boys" brought to my day and life was unsurpassed! My reasons for fighting the daily battles were here and everything was going to be just fine! Jeff was always a hit with all the nurses and he loved to ride up and down the hall on my lap in the wheelchair.

We played, talked, and just spent time together until the evening grew late and Jeff had to get home to bed. How I missed putting my little boy to bed and the family time we always had with him. Hopefully, I would be home soon and out of the wheelchair able to do motherly things again.

The following day was a kidney dialysis day. Every Tuesday, Thursday, and Sunday I ran on the machine for ten hours. Today I would teach the nurses how to put me on the machine. They were unfamiliar with kidney machines, so I would show them what to do and how to do it. I had learned much since our training at Good Samaritan Hospital.

Everything ran smoothly, but it was a relief when Jerry stopped by to see how things were going. How I depended upon him! He was my "Rock of Gibraltar," and I knew he could handle almost any situation that might arise. He gave me a sense of peace whenever he was near and it made running the kidney machine seem all right.

Wednesday finally arrived and it was to be an exciting, active day: first to vocational rehabilitation, then a meeting with Dr. Drake and lastly, physical therapy. At nine-thirty an orderly arrived in my room to wheel me to the vocational rehab department.

"Good morning," he said. "Are you ready to go for a little ride?"

"Sure thing," I answered, and off we went down the long, sterile hallway to the elevators. He pressed the "down" button and we waited like statues staring at the numbers above the door as they lit up at each floor. Finally, the third floor's number was up and the bell rang, signaling that the elevator had arrived. As we entered the door, I had a self-conscious feeling about being the only one sitting down. People towering above always seemed to gawk uneasily at a wheelchair person. I felt equally uneasy and knew they were probably wondering what had happened to me.

This feeling was easy for me to surmise, because not too many days ago, I had been a "standing up" person on an elevator when someone had entered in a wheelchair. Despite the uneasiness I felt, it was going to be necessary to get used to these feelings—people staring, little kids pointing, gentlemen holding doors, and curious glances all around me. Get used to it—that is—for just a little while. I hoped it wouldn't be long before I would be out of this thing and back to normal.

"Coming out, please," my escort called from the back of the elevator. We had reached the basement level where the vocational rehab department was located. As he wheeled me into the room, he said that he would be back for me in about an hour and a half. I thanked him and looked up to see the therapist standing in front of me.

"Good morning, Ann. My name is Miss Betty. I'm here to help you get started on a project today. I understand you need something to help strengthen your hands. What do you like to do? Paint? Clay work? Wood work? Copper tooling?" she suggested.

"What is copper tooling?" I inquired.

"I'll show you," she replied. "Over here in this box I have some roosters that need to be cut out with a saw and then we put copper feathers on them. Then they are attached to the wood. We hang pot holders on them. Do you think you'd like to make one?"

"Yes, I would. They really look neat and it would take quite a bit of hand and arm muscles."

"Right," she agreed. "I think that it would be a good project for you. Come over here, and we'll get started."

Other people had come in by now and were sitting around the table working on projects they had started earlier. One of the fellows from my floor was down there. His name was Scott, a quadraplegic who was learning to type by hitting the keys with a pencil held tightly in his teeth. I said a little thank you prayer silently to God. Although my hands were very weak I could still use them, and they had a good chance of getting better and stronger. Scott didn't have much of a chance for improvement. An automobile accident had left his limbs virtually lifeless. Praise the Lord that he gives us no more than we can bear. I knew that somehow, some way, I could cope with kidney machines and wheelchairs, but didn't think I could have coped with Scott's problems. For the present I didn't have to, and I thanked God.

During my stay of about five weeks in the hospital, the vocational rehab department filled many hours for me. People and projects helped me to gain strength in my fingers and hands, which had been slowly wasting away, and to gain strength in my soul. Many patients in the department were much more handicapped than I, and seeing their determination and success cultivated my inner desire to try harder and to be a winner. With God's help, I would be!

After my rehabilitation session, my escort arrived to wheel me to my appointment with Dr. Drake. When we arrived, we entered a room equipped with two long tables and a few chairs. Dr. Drake was not there yet, so I rolled up to the end of one table to wait. My escort gave me a magazine to browse through and said he would return in about half an hour to take me to physical therapy. What a busy afternoon, but I was glad for it. Before long, Dr. Drake arrived.

"Hello, Ann," he said. "How are you doing?"

"Okay," I replied. "Did you know that I'm in this wheelchair all of the time now? Dr. McMahon has me in the hospital to watch me closer and to give me therapy also."

"Yes, I know that. He has been keeping in close contact with me and is doing a fine job. The main reason I have come over here today is to tell you that the dialysis department is going to be managed by another doctor. I have been called to serve in the armed service and I will be leaving here soon," he said.

"Oh, my goodness," I replied in surprise. "How long will you be gone?"

"Two years," he answered. "I'm sure you will like my partner, Dr. Martinson. He is very competent and will oversee you just as I would. Your next appointment will be with him."

"Okay," was my stunned reply.

"Now tell me about you," he continued. "How's Jerry and what have you been doing here at Emanuel?"

"Jerry is fine," I informed him. "This whole thing is hard on him, I know, and now that I'm in a wheelchair, it's even harder. He has Jeff to take care of, the house and everything. I can see that illness is not easy on our marriage."

"No, it isn't," Dr. Drake agreed. "You seem to be a strong couple, though, and you'll see it through. Just let Dr. Martinson know if he can help with anything."

"I will," I said.

"Now, what about the hospital. What are they having you do?" he continued.

"Well, I start physical therapy this afternoon, and I started occupational therapy this morning. My legs and hands are so weak, and I guess they are trying to strengthen my muscles."

"I'm sure you will get along just fine," he assured me. "It's really nice to see you, but I must be going. I wish you the best of luck and health."

"Thank you and good luck to you, too," I replied. "I really appreciate all that you have done for me."

"That's what I'm here for," he answered. "Goodbye."

As he left the conference room that afternoon, I sensed that I would probably never see him again and I'm sure he thought this was his last visit with me. At the rate I was going, I didn't know if I would even see Christmas, and it was only a few weeks away!

My escort returned for me shortly and we were off down the hall and waiting for an elevator to take us to the physical therapy department.

The therapy department was a busy place and I was greeted by the head therapist.

"Good afternoon. My name is Tom," he said, introducing himself.

"I'm Ann Van Winkle."

"I'm glad to meet you," said Tom.

He took my chart from the orderly and began looking through it.

"Looks like your main problem is your lower legs. Is that right?" he questioned.

"Yes," I answered, "and my hands."

Okay," he replied. "Sue will be working with you today. She'll be getting you out of your chair and onto the mats where she will exercise your feet and legs. Then she may have you work on some of the equipment we have. First we will start you out on the tilt board. This board gradually tilts you to a vertical position while you are strapped to it. In this way, you get used to standing up again. We will put you into the whirlpool tubs and eventually you will be riding the stationary bicycle, lifting weights, and working on some other equipment to strengthen your muscles. Does that sound okay?"

"Sure," I replied. "I'm ready to do anything to get back on my legs again. I'm tired of wheeling myself around!"

Just then, Sue entered the room and Tom introduced us.

"I'm glad to meet you," she smiled. "Are you ready for a workout?"

"Sure am. Let's go!"

She wheeled me into the physical therapy room and helped me down onto the mat. My muscle ability was tested by asking me to move my toes (no response), twist my feet in and out (no response), raise my foot up, pulling from the ankle (no response), and push down toward the ground (some response). In all of the tests, my muscles were very weak if there was any muscle at all. She continued to move my feet and legs and gave me exercises to do on my own.

"Roll over onto your stomach," she said. "Now push your feet up toward the ceiling by pushing from your ankles."

I got a B + on that one and she decided to have me lift weights with my legs for a while. Then it was on to the whirlpool, and I let my body completely relax in this warm, soothing vat of whirling waters.

After many sessions on this equipment, I graduated to riding the stationary bicycle. Two therapists had to lift me onto the seat and then Sue stood right by me to make sure I didn't fall off. My goal at first was to ride one-half mile and this I did without too much difficulty.

After several days on the mats and the leg strengthening equipment in the therapy room, my therapist decided I needed practice walking. Because I was scared to death of standing up since I had fallen so much, she knew it would be asking too much to have me walk with the aid of leg braces and crutches just yet. Instead, she came up with the brilliant idea of having me walk in the water. The buoyancy of the water would hold me up and I couldn't fall. She could teach me to walk all over again without the fear of falling. It sounded like a wonderful idea to me. I loved being in the water and felt confident there, which was very important at this point.

Luckily, Emanuel Hospital had a beautiful swimming pool that would facilitate this new endeavor. Sue had some handrails put into the water and I could walk between them, holding on as I gingerly picked up one lifeless, numb leg, set it down, and repeated the procedure with the other leg.

The first day in the pool was very exciting. As Steve, another therapist, wheeled me into the pool area, a feeling of exhuberant peace came over me. Finally, in here, I would be free. I knew I could swim and could hardly wait.

"How am I going to get in there?" I asked.

"Don't you worry," Steve reassured me. "I'm just going to pick you up and carry you in there."

As I was giggling that's what he did—just picked me up and placed me in a floating position in the water.

"Are you all right?" he questioned.

"Fine," I answered. "Just fine! Just let me go. I know I can swim."

As he gently took his hands out from under me, I experienced the freedom I so much wanted. I began paddling my way to other parts of the pool, and the feeling was fantastic. I had almost forgotten how good it felt to be free. In here, I was not dependent upon wheels to get around. I could maneuver my own body and go wherever I wanted to go. It was great!

Sue and Steve came in and swam with me and it was a strengthening, good experience. I felt as though we were on an even plane. In here, for a short time, I was no longer the patient, but we were friends who were just having a good time in the water.

Finally, we had to get down to the task at hand—

learning to walk again. It would be so much easier, I thought, if the whole world were a swimming pool. I could manage that and not fall down. Getting back to reality, however, I swam over to the bars. Back and forth I walked, lifting my legs and exercising the muscles that once held my legs erect.

After many steps along the same water path, it was time to get out. Steve once again picked me up from a floating position and carried me out of the water, placing me gently in the wheelchair I called my own. Back to the world of stares and wondering eyes as we made our way through the hospital corridors to my room.

After two weeks of exercising in the pool, Sue decided that it was time for me to get fitted for a pair of leg braces and to start trying to stand on my legs and walking a little with the help of crutches— on land this time.

The "brace man" was called to come and fit me with a pair of leg braces. Leg braces did not come in a variety of styles or colors. I was fitted with the only type available. They were made of thick steel that fastened to an oxford-type shoe. The steel had a hinge at the ankle that bowed out a bit. Two steel supports, about one-inch wide, ran from the bottom of the shoe, up each side of my leg and ended just below my knee. They fastened together at the top with some velcro fasteners. The braces were very heavy, but held my feet at ninety-degree angles from the leg which held my "drop foot" up so I didn't trip over my toes when I walked. When I attempted to walk without braces, I had to lift my leg very high to clear my drop foot from scraping my toes on the ground and tripping.

The braces felt very secure when I put them on. Handing me a pair of crutches, Sue said that today

I was just going to try standing—no walking. She explained that I was to put my weight on the hand bars of the crutches and not on the part under the arm.

"Okay," she said. "Are you ready?"

"I guess so."

"Now, I want you to hold your crutches with one hand and push up on the wheelchair armrest with the other. Be sure your wheelchair wheels are locked. Balance yourself when you get up and then transfer one crutch to the other hand. Don't be afraid. I'll be standing right here beside you and I won't let you fall. Your legs are a lot stronger now from working in the pool and you can do it."

"All right, here goes," I said. "I'm afraid I'll fall when I get on my legs."

"No, I won't let you fall," Sue promised.

I was so afraid. I had ended up in a heap on the floor so many times that I had a genuine fear of standing up on my legs. But, somehow, some way, I would overcome.

"Come on," Sue repeated, "let's try it."

So, with my heart pounding, knees quivering, and the palms of my hands clammy with perspiration, I stood up. Hallelujah! I felt like a babe who wanted to cheer! I wanted to tell the whole world that I, Ann Van Winkle, was standing on my own two legs and I was not falling down. What an accomplishment! I felt joy throughout my entire being. I was up on my feet, but no steps were taken that day.

I went back to my room, feeling like I had really accomplished something. I could hardly wait for the next therapy session, which was forty-eight hours away. Tomorrow would bring another session on the kidney machine, so there would be no therapy.

During my next therapy session, Sue decided to

put a piece of tape on the floor as the goal to which she would like me to walk. So after my initial standing performance, she said, "Now today I just want you to take one step away from your wheelchair. Then, we'll turn around and go back."

"What a chore," I thought.

As I looked down the long insurmountable hallway, it seemed impossible that I would ever walk that far! It seemed twice as long as a football field. How could I ever do it? However, one therapy day at a time and one step at a time, I walked one additional linoleum tile square each day. The day I got halfway down the hall was really a day of victory. Determination pushed me on!

Halfway down the hall I walked, three-fourths of the way down the hall, and finally, the sense of real accomplishment when I reached the end of the long, challenging hallway. I felt triumphant, but I knew in my heart I still had many mountains to climb. However, Sue and I stood and cheered and were joined by some passersby.

"I didn't know if I would ever get down here!" I exclaimed. "Guess I really did it!"

"You sure did," Sue said excitedly. "I knew you would. Before long, you'll be out of that chair for good!"

"Oh, do you really think so?" I asked hopefully.

"Sure you will," she replied. "With your determination I have no doubt in my mind that you could climb mountains."

"Climb mountains," I thought. "Wouldn't that be neat. What a dream!"

But coming back to the job in front of me, or behind me, I should say, I had to turn around and walk back down the long narrow hallway to my wheelchair. So, we started out. When we reached

the wheelchair, I sat down, tired and weary, but full
of joy. I had done it! Now, there had to be another
challenge.

Sue spoke, "Tomorrow we'll go outside and walk.
We'll start learning how to go up and down curbs
and steps. We have a lot more to accomplish before
you go home. We'll do it together."

"Okay," I replied. "It sounds like a real challenge."

The tomorrow we spoke about ended up being
another day on the kidney machine, but the next
day came soon. The day was bright with winter sun-
shine and just right for a walk outside.

By this time, I was wheeling myself down to the
therapy room and Sue met me in the hallway.

"Are you ready to go outside?" she asked.

"I guess so," I answered with a bit of hesitation.

"Let's go into the therapy room and work on the
equipment first," she said. "Then we'll go walking
outside."

We went over to the leg weight table where Sue
helped me stand up and boost myself onto the table.
Five pounds more weight was added to my weights
today, making a total of fifteen pounds to lift. With
every leg lift, I thought, "Muscles get strong, muscles
get strong." Then, after the leg weights, came the
exercises on the mat and then the bicycle. Three-
and-one-half miles was my goal today.

As Sue worked with me, we talked.

"How are your new leg braces?" she asked. "Any
problems with them?"

"No, not really," I answered. "It's just that they
are heavy and awkward, but I guess I'm getting used
to that. I was always one to love shoes, too. But these
oxfords, laces and ties, are not particularly stylish.
They make me think of old ladies with bun-style
hairdos. Guess I shouldn't complain and I don't mean

to, but I guess I just have a lot of things to get used to."

"You'll do it," Sue assured me. "And besides, I understand you can get those shoes in a variety of colors."

We both laughed and the humor of her comment rang out a note of joy.

"Ready to go outside?" she inquired.

"Ready or not," I said, "let's go."

This time Sue wheeled me to the elevator and we made our way through the long hallways and out a side door. Sidewalks and curbs were right before us. How open, free, and somewhat frightening it felt to be outside. There was so much to conquer out there—much more to conquer than a single swimming pool, long seemingly-endless hallways, and elevators. This was the real outside world with many obstacles, and the first one before me was a curb.

"Here are your crutches," Sue said. "Now, I want you to stand up just like you've done downstairs and balance on your crutches. Ready?"

"Okay, ready," I replied.

I pushed myself up from the wheelchair, got balanced on the crutches and began to take a few steps.

"How does it feel?" Sue asked.

"Great, but I'm a little shaky," I answered.

"That's okay," she added. "Now, walk over here to this curb."

At a snail's pace, I approached the curb.

"Okay, now put both crutch tips onto the street," she directed.

I did it.

"Now, one leg down and then the other," she continued.

This I did also, moving one foot at a time to get my balance.

"Perfect!" Sue exclaimed. "Now, turn around and we'll go back up. Okay, now both crutches in the street. Lift one leg at a time onto the curb, push with your hands holding tightly to your crutch hand bars, and finally, bring the crutches up on the curb."

Following her instructions, the feat was accomplished without much trauma. Maybe, just maybe, this outside world wasn't so bad after all.

"Push on, keep going," I told myself. Some day maybe this outside world would be mine again. I'd give it a good try.

We can rejoice, too, when we run into problems and trials for we know that they are good for us—they help us learn to be patient.

—Romans 5:3

Chapter 8

Emergency!

"Ann Van Winkle to see Dr. Martinson," I said as I reported in at the front desk of the office formerly occupied by Dr. Drake.

Since Dr. Drake had been beckoned by Uncle Sam, Dr. Martinson, his partner, had taken over my case. Dr. McMahon no longer felt knowledgeable enough about kidney problems, so he sent me back to my original team of doctors after I left Emanuel Hospital.

"Oh, yes, Ann. He'll be right with you," the receptionist answered.

I was there this bright May morning with not so bright an outlook on life. I was feeling very weak, tired, and unmotivated. My stools had been turning black in color and the doctors were suspicious that I might be bleeding internally.

"Come in, Ann," beckoned Dr. Martinson as he stood in the doorway of the waiting room.

"Step on the scales, please," he requested. "Um-m, been losing a little weight it looks like."

"Probably so," I replied. "I just haven't felt like eating very much."

As we walked into his office, he glanced through my chart that was on the desk before him.

"Now, tell me, how have you been feeling?"

"Not too well," I confessed. "It seems like everything is a big, huge effort. I don't have much pep. I'm not hungry and I just don't feel good."

"Okay," he replied. "I want to admit you to the hospital this afternoon. When is your next kidney run?"

"Tomorrow," I answered.

"Fine," he said. "We'll run you at the center after you've had a good night's sleep. Can you check in by about four in the afternoon?"

"Sure," I replied.

In my mind I was both happy and sad that another hospital admittance was right around the corner.

It was during this time that cloudy, dismal thoughts again began entering my mind. What was in store for me tomorrow, the next day, the next week, month, and perhaps next year? Would the next Christmas ever come for me with family, friends and all of the holiday festivities that I loved so much? Would I ever see little Jeff enter the first grade or graduate from high school? The uncertainty of it all would almost get the best of me when my favorite Scripture verse would step in and take over:

> And we know that all things work together
> for good to them that love God and they are
> called according to his purpose.
> —Romans 8:28 (KJV)

"Yes, but do I really love Him enough? Maybe that verse is really just for someone else," I thought.

As I packed my suitcase for the hospital, thoughts ran through my mind that maybe this would be the last time I had to pack it. How could things get much worse? What I didn't know at that point was God still hadn't given me as much as He knew I could bear and there was more to come.

We routinely checked into the dialysis unit and I was soon hooked up to the kidney machine for my needed ten-hour dialysis run. The next thing I remember was that I felt terrible. Really terrible! I told Jeannie, my nurse, that I thought I was going to throw up, and throw up I did. But, this was different. This time, I vomited blood. Jeannie knew this was a life and death situation. The Heparin that I had been given when going on the machine had thinned my blood to the point where it would not clot. Obviously, I was bleeding internally, and if it wasn't stopped soon I could bleed to death in a relatively short time.

"Paging Dr. Hayes, Dr. John Hayes," the voice sounded over the hospital address system.

Where was he? Had he already left the hospital? Jeannie waited anxiously on the other end of the telephone line.

After what seemed like an eternity, a call came in to the hospital operator.

"This is Dr. Hayes," a voice responded.

"Dr. Hayes, you are wanted for an emergency in the dialysis room. Please dial extension 412."

Jeannie quickly answered his response.

"Dr. Hayes, Ann Van Winkle is on dialysis and has just vomited quite a bit of blood. We are suspicious that she is bleeding internally. We need immediate orders as to what to do, please."

"Stop the Heparin immediately and take her off dialysis. I'll be right over," he replied.

The orders were carried out by the time he walked through the doorway into my room. His large, powerful but gentle stature had a calming effect on me. If Dr. Hayes was present, some way, no matter how sick I was, I knew everything was going to be all right.

Upon examining me very quickly but thoroughly, his suspicions were that I had a bleeding stomach ulcer and that immediate emergency surgery was necessary.

"Prepare her for surgery, immediately," he said. "She's too weak to go over in a wheelchair so we'll have to take her on a stretcher. Don't get one with wheels because it won't ever fit in that old elevator. We'll carry her down the stairs and over to surgery. Administer ten milligrams of protamine, I.V., to help control the bleeding," he added.

"Yes," I thought, "eveything is going to be all right. Dr. Hayes is here and he is in control. Everything will be just fine." I settled back to pray that God would guide the nurses and doctors in their efforts to help me.

The dialysis center was bustling with emergency activity. I was carefully loaded onto the stretcher, strapped tightly to its metal frame and carried down the steep staircase leading from the dialysis unit, out across the street and into the main hospital where a surgery room was being prepared for me.

The trip down the stairs was not too bad until they came to the corner where the stairs turned sharply at a ninety-degree angle. Somehow, Dr. Hayes and Jeannie managed to get the six-foot frame around the corner and we were at the surgery room door in a matter of minutes.

"Ann Van Winkle for emergency surgery," Jeannie stated. "Doctor's orders are to prepare the abdominal area for surgery, immediately."

Dr. Hayes was scrubbing down in preparation for

surgery as I was transferred from the carrying frame to the surgery table. I was prepared very quickly and also given a blood transfusion.

"Blood pressure reading, please," said Dr. Hayes as he entered the surgery room.

"100 over 64," replied the surgery nurse.

"Keep saline running at 200 cc.'s an hour and give one unit of whole blood," Dr. Hayes stated.

"Yes, Doctor," said the nurse.

"Mrs. Van Winkle, I'm Dr. Wheeler," said a pleasant voice at my right side. "I'm going to be your anesthesiologist. In a few miinutes we will be inserting a small needle into your vein to administer Sodium Pentothal. You will go to sleep and when you wake up you will be in the recovery room."

He was right. I do remember a little stick and then he asked me to count to ten. I remember getting to three and then the next thing I recall is waking up in what I thought was the recovery room. I was sore and felt very weak!

"Mrs. Van Winkle, you're going to be all right. Your surgery went well. We'll be taking care of you in here for a while and we'd like you to just lie back and get some rest."

"Where am I?" I questioned.

"You're in the intensive care unit. Dr. Hayes just operated on your stomach and you are doing just fine. Your husband will be in to see you in a few minutes."

I discovered later that two-thirds of my stomach had been removed. As I lay back on my pillow, I could feel bandages pressing against my very sore and tender stomach area. Tubes were coming out of my nose and both arms and I reconciled myself to the fact very quickly that I wasn't going anywhere. What's more, I didn't *feel* like going anywhere. I was completely obliged to let someone else take care of me for the time being.

"Hi, honey," a familiar voice said. "How do you feel?"

"Sore," I replied.

"Well, just rest. You're going to be just fine."

It was my strong, understanding husband who was by my side, and his presence made everything seem all right.

The days in intensive care went by faster than I had expected. I spent most of my time sleeping and occasionally watching the television set that was perched on a shelf, high up in the corner of my tiny room. It wasn't long before I was transferred to a regular hospital room where the nurses began getting me out of bed, onto my wobbly legs and walking a little.

It was an awkward position that I assumed when standing because my stomach hurt so much. I felt like a hunchback as I walked, bent over, to protect my middle. It was so sore!

Jerry came to visit me one afternoon, announcing that he was looking at a travel trailer he thought he would buy. As soon as I was able, he wanted to take me out in it for the weekend. He would ask our good friends Toni and Tony (Toni girl and Tony boy, we called them) to go along with us and help take care of Jeff. It sounded like a great idea, but a little frightening.

As I again adjusted to hospital life, I found myself becoming quite accustomed to the security around me. It would be difficult to leave. However, I knew as soon as my food intake became better, I would be going home.

Eating was a very slow process. The one-third of my stomach that was left was not tolerant of too much food entering it at one time. Dr. Hayes had instructed me to eat six times per day until the stomach stretched out again. Amazingly enough, I learned that it would stretch out to its full size again.

After a few more days in the hospital, they sent me home to be independent again, returning to the routine of running on the kidney machine by myself.

It was neat to finally be home. Little Jeff was very caring and very concerned about Mommy's "outchie" and all of the bandages that surrounded my stomach. Mrs. Seeley, my housekeeper, and Jerry took good care of me, and within two weeks Jerry decided that a weekend trip in the new trailer was in order. So we ventured out to the Mount Hood area to camp with Toni and Tony for the weekend.

Everything went super during the weekend and our lives seemed to be calming down a little. We were getting back into the swing of family life that was molded around an artificial kidney machine when emergency number two struck.

Six weeks after my stomach surgery I was back at the doctor's office. This time, the problem was my calcium. All blood tests indicated that it was too high. The count was number twelve on the calcium scale, and I was told if it rose much higher, my heart could stop beating. The immediate answer was to go back into surgery and have some little glands called *parathyroid glands* removed. They are small glands located in the neck which control the calcium that is used by the body.

"Okay," I told Dr. Martinson, who was heading up this venture. "I guess at this point, I don't have much choice. Do or die? Right?"

"Well, Ann, that's putting it bluntly, but you're right."

The path to the surgery room was well worn by me now. I knew the procedure well. Too well, indeed. My arms and now my stomach and sides were road maps that I would later refer to as my "battle scars." This time, however, I went into the entire procedure somewhat dubious, for my body had not yet recovered from the last battle.

The hour of surgery came and I was prepared as usual. By this time, the nurses knew I didn't want anything to relax me in the room before my journey down the long, sterile hallway and up the elevator to the surgery floor. I preferred to have all of my senses and to know what was going on.

Dr. Hayes performed the surgery and everything went well. I remember waking up in the intensive care unit again. It was becoming harder and harder to breathe. My throat, where the operation was performed, was very sore but this difficulty in breathing didn't seem quite right. My head was not clear. I wasn't thinking straight and I remember telling myself not to panic. I rolled over in search of the bell to call the nurse. Panic was really setting in now.

"Why can't I breathe freely?" I said to myself.

The palms of my hands were beginning to sweat. My heart was beating rapidly and it became increasingly hard to think.

What's happening? Please, somebody come!

Just as I rang the bedside bell, a nurse popped her head in the door.

"How are you doing?" she questioned.

"My throat!" I gasped.

In a flash she was out of my room and on the emergency line to Dr. Hayes. She sent another nurse in to keep me calm.

"Dr. Hayes, this is Miss Stenson in intensive care. Mrs. Van Winkle is having difficulty breathing," she said. And, before she ever had a chance to continue, Dr. Hayes replied.

"I'll be right there!"

And that he was.

"Tracheotomy tray, please," he said as he quickly examined my throat area.

"It's right here in the drawer, Doctor," the nurse replied.

The tray had been ordered earlier just in case there was an emergency such as this.

"Good. Sterile gloves, please," he said.

And then, leaning over my bed he calmly said, "Ann, you're going to be all right in a second. This is going to hurt for just a minute."

At that point, I almost lost consciousness but I recall the sharp point of a knife coming down on my throat and then...I was out. Dr. Hayes had opened an air passage for me by cutting into my windpipe while I was lying right there in my bed. When I awoke, there was a tube sticking out of my throat and it felt funny to breathe. The nurse informed me that I had swollen up so much from the parathyroid operation that it cut off my oxygen. It was necessary to insert a trachea into my throat so that I could breathe.

God, in His infinite wisdom, somehow reaches down and lets us know that even through traumatic trials and sufferings He does care about us and loves us to the fullest. It is most difficult during these periods in life to look upon this as love. But like fine gold, some have to be put through the test of fire before they come out solid and pure. And, through it all we are to:

> Love the Lord your God with all your heart and with all your soul and with all your strength.
>
> —Deuteronomy 6:5 (NIV)

I called upon your name, O Lord, from deep within the well, and you heard me! You listened to my pleading; you heard my weeping! Yes, you came at my despairing cry and told me not to fear.

—Lamentations 3:55-57

Chapter 9

A Way of Life

Surviving became an everyday goal after leaving the hospital. We rejoiced and gave thanks to God when we were back home and on our own.

The storm winds of life were still churning, however, because it was not long until hepatitis, a liver disease, stepped in to take its toll on my already degenerated body. People die from having hepatitis alone, so combined with my other problems, this new disease made the future look even more bleak. Life was a kaleidoscope made of gray and black chunks of glass. I never knew how many or when the black pieces would fall into place.

Once again, however, we weathered another difficult situation, plodding wearily into the future.

Days, months, and years slipped by unrehearsed. Like birthdays that creep upon us when we are not looking, our lives were somehow being lived.

Running on a kidney machine became such a ritual that it was a naturally accepted part of our living routine. Going on the machine three nights

103

a week for ten hours had become a chore much like washing clothes or doing the dishes after dinner.

Jeff was especially accepting of this situation. He became used to "helping" Daddy set up the artificial kidney machine in preparation for "Mommy's run." He knew that if he was very speedy and put his own pajamas on, brushed his teeth and got all ready for bed, that just maybe he could help Daddy hook Mommy up to the machine. He also loved to have his blood pressure taken, and he knew that if he was around when Daddy took Mommy's blood pressure, he would probably get his taken too. Such fun! He also got to listen through the stethoscope. It was a highlight of his day to hear his heart go "burumph... burumph...burumph."

After getting on the machine, if there was time, I would often read Jeff his bedtime story before Jerry tucked him into bed for the night. This was a special time for both of us.

So Jeff, at his young age, was growing up not knowing there was any other way to live. We didn't realize this until one day when we visited some friends.

After we had arrived and greeted everyone, Jeff became restless and began exploring the house with the inquisitiveness of a three-year-old. He poked his small head into the bedroom and hurried back into the living room where we were all engaged in conversation.

"Mommy, Mommy!" he exclaimed. "Where's their kidney machine?"

Everyone laughed and poor Jeff didn't know why. At his young age, he thought that all moms went on kidney machines at night and these people didn't have one. As you can imagine, that took some explaining.

"Jeff," Daddy said, "they don't have a kidney machine. The mom in this house doesn't need one. Our mom is special. She is the only mom we know who needs one. And, you know what?"

"What?" Jeff asked.

"Maybe some day even your mom won't need a kidney machine! Wouldn't that be great?"

With that, Jeff lost interest in the entire subject. It was too much for his young mind to grasp. Actually, it was more than even my adult mind could easily comprehend.

●   ●   ●

One day I decided to go for a little walk outside by myself. Hanging tightly to just one crutch to balance myself, I walked across the street to our mailbox. It was a crisp autumn day and leaves were being blown gently ahead of me. The tip of my crutch scrunched them as I walked. The colors of the season were magnificently marked in the red maple leaves as they floated in the gentle breeze and in the yellow and green leaves from a nearby sycamore tree.

As I walked along, I recalled episodes I had experienced during the past few years. I chuckled when I remembered a young newspaper reporter who asked me, using the current slang, how I got "into" the kidney machine!

I thought about the heavy steel leg braces that were secure around my lower legs. I recently had heard about some new spring-wire braces which the doctors felt I might be ready for. They did not offer as much support, but they felt these newer braces would be adequate for my leg problems. There was still a glimmer of hope in my mind that some day I would be able to regain the movement in my lower legs. Maybe these new braces were a positive step so I was anxious for next week. The brace shop would fit me and then I would be able to have a little wider choice of shoes to wear. That would be great because my wardrobe of footwear had been limited to two pair of oxford-type, lace shoes.

The new braces proved to be very satisfactory and I loved the idea of more shoes in my wardrobe. Now I had three different kinds and colors—all with straps to hold them on instead of laces.

Time progressed and my kidney "log" that was kept of each dialysis run began to number runs in the 900s. Three runs a week with fifty-two weeks in a year added up to 156 runs each year. It wouldn't be long before I would hit number 1000. It was my personal goal to have some kind of a celebration when I achieved that number. I didn't know quite what, but we had to do something extra special. Other people had things in mind also.

By this time, the Drake-Willock Company had become a large corporation. Much growth had occurred since Charlie Willock and Dr. Drake had invented the first home dialysis machine for Mark Smith back in the fall of 1962. Even though it was a large company now, they were still a very personal company and became intrigued with my projected 1000th-run record on one of their machines. Sally Ryan, who worked for the company (and whose mother was also on dialysis) began research on just what 1000 kidney runs meant:

1000 TREATMENTS
= 10,000 HOURS

= 24 hours a day for 413 days—working 7 days a week for 3 years (without pay)

= time enough for 382 round-trips to Paris, France...

or

time enough for a leisurely walk around the world.

Her blood has traveled 3,247 miles outside the body, with the help of 79,218,396 heartbeats.

WOW! was the comment heard from many people.

Along with printing my success story in their company newsletter, the entire company plus my doctors, family, and the Kidney Association of Oregon staff turned out for a celebration of my 1000 runs on the machine. Champagne was hung to mimic a bottle of saline hanging on a rod above the kidney machine. And the goldfish? No one knew how it got inside the champagne bottle! Newspaper photographers and reporters covered the event and goodies were served to all.

Ironically, my cannula started acting up about two days before the event. Dr. Hayes had to operate on it at that time so I didn't know if I'd be able to attend the party. Almost nothing would keep me from going, however. Even if I had to get back in my wheelchair, I *would* go. Thankfully, I didn't need to go in a wheelchair but it was a near possibility. The excitement of the event kept me going and it turned out to be a splendid day!

Yes, the machine had certainly become a way of life for my entire family.

•   •   •

"What does the future hold?" I asked Dr. Drake as I sat in his office one bright morning. It was so

good to have him back now from military service.
I knew he would really tell me the truth.

"How does the transplant situation look today? I
see some of my friends who are doing pretty well
with their transplanted kidney and I am wondering if that is for me." Millie, an old friend who
started training on the machine when I did, had
gone off the machine, opting for a kidney from a
relative. It was an "A" match but she rejected it in
six weeks. Another kidney replaced the rejected one
(this time from a cadaver) and she had retained it
for about four years.

Then there was Sue. She was doing well with her
transplanted kidney, too, and so were others I had
known on the machine.

"What do you think my chances are?" I questioned Dr. Drake.

"To tell you the truth, Ann, your chances at this
point on the machine are about the same as they are
with a transplant. It is a gamble, though. You know
that. You are doing fine on the machine and will
probably continue that way for some time. However, if you want to go talk to the University of
Oregon Health Sciences Center about a transplant,
I am sure you would feel better and lead a more normal life if you were suited for a transplant. I would
suggest that you go talk to Dr. Lawson, the head
kidney transplant surgeon at the center, and see
what he says. They can't do any more than turn you
down. Talk it over with your husband and let me
know tomorrow what you decide about having a
transplant or going on the way you have been."

I left his office weighed down with the knowledge
that my future depended upon this decision. That
evening, Jerry and I probed the possibility of a
transplant versus staying on the kidney machine.

We discussed not only the physical aspects of the machine but also the social and emotional aspects we were going through.

To begin with, there had been the touch-and-go moments of life and death. Then came the wheelchair when my legs wouldn't function. Now it was crutches and perhaps a cane some day. There had been lots and lots of pain from bones disintegrating in my back. I was forced into wearing an awkward back brace for two years until Dr. Drake put me on Hytakerol, which did wonders. The pain was gone within three months. This medicine allowed the calcium which was being robbed from my bones to be better absorbed by my system.

As a family, we were regulated to the machine three nights a week, which meant we had to plan things outside the home on alternate nights. The machine always came first.

It was difficult to do my own housework so I needed physical help from other people.

The diet I was on really affected the whole family. Everything was cooked without salt with Jerry and Jeff salting at the table. It was very difficult to go out to eat—especially in the fast-food restaurants. Pizza was definitely off-limits and commercial hamburgers were usually salted when they were cooked.

In our marriage relationship, things were not smooth. When I was not feeling up to par, our sex life really went downhill. This was an area that needed lots of special consideration. For my part, I felt like I needed to give a lot even if I didn't feel good. We both felt that this physical area could make or break a marriage, so we both worked hard to keep harmony.

Socially, we didn't have time or energy for much except seeing old friends occasionally. Running on the machine thirty hours a week and then re-building it took most of our time. Thank goodness my brother, John, and others helped me by doing this when Jerry was out of town on business.

It was difficult, but not impossible, to take short vacations. We would generally leave on a Friday and be gone until Sunday afternoon. We took our travel trailer and toured many parts of Oregon. We would always be back by Sunday evening to run on the machine.

One summer we took an extensive trip into Canada, but I had to plan ahead and run on machines in hospitals in Seattle and Vancouver, B.C., while we were gone.

People usually accepted me the way I was. Little children would often stare at me in the wheelchair or look with questioning eyes at my leg braces. On the whole, people were warm and giving.

The emotional aspect of kidney disease is pro-bably the hardest we felt. Many marriages fail during traumatic illnesses because one or both marriage partners cannot cope with the physical and emotional changes that take place during this time. Because both people are uncertain of the future, it is very difficult to communicate. I can remember being very emotional about my child most of all. I wanted to live long enough to see him grow up, plus I hoped he would accept my plight in life.

Most of the time a positive attitude kept me look-ing at the bright side of things, and Jerry and I kept pushing forward together. "Livin' " was our motto and we stretched the clock as much as possible and made the most of every day. Yes, in fact, every minute of every day!

What I do today is important
because I am exchanging a day
of my life for it.
                    —Author Unknown

Statistics always tried to crowd my positive feelings into a corner. Statistics claimed a person could
live on a kidney machine two to five years. Statistics
said that the person who had lived the longest on
a machine had lived twelve years on it. Statistics
became a reality when I saw people who died after
living on the machine for a short year or sometimes
two or three years.

Unbeknown to him, Billy Graham was a great in-
fluence on this part of our lives. As a young child,
I attended Sunday school and church regularly with
my family and this was always a very important
part of my life. I grew up in the church, but never
really knew what it meant to turn my entire life
over to God and to trust Him—no matter what!

Seek ye first the kingdom of God, and his
righteousness; and all these things shall
be added unto you.
                    —Matthew 6:33 (KJV)

I knew in my heart that I needed to seek God's
kingdom first and to give my life to Him. Then God
would take care of everything else. Being an "I'd
rather do it myself" person all of my life, it was difficult to say, "Okay, God, I know I am a sinner. I
believe that Jesus died to save me from those sins
and I ask you to take control of my life—no matter
what—and please live in me daily."

Yes, it was hard to "let go and let God" but on a
special night in May, 1968, Billy Graham was in
Portland so Jerry and I decided to go hear him.

Upon arriving, one of the ushers quickly whisked me off in my wheelchair to the lower level and put me right by the stage where Dr. Graham was to speak. He went into an adjacent office and brought a folding chair for Jerry to sit on since all the others had been taken. It was very plain to see that God wanted us here!

We sat through the beautiful music and Dr. Graham's powerful message. He said there were many people in the audience that needed God to control their lives, and all we needed to do was to ask Him.

After bowing in prayer, Dr. Graham said, "Just come, come as you are—men, women, young, old, church member, church deacon, whoever you are— just come."

Jerry looked at me and I at him. He said, "Do you want to go? I do."

"Yes," I answered, and we both went forward to proclaim Christ the Lord of our lives.

> Trust in the Lord with all
> your heart and lean not on
> your own understanding.
> —Proverbs 3:5 (NIV)

This verse would become my "rock." Jerry would help pick me up emotionally when I would stumble in the weeks and months ahead. I am sure it must have been extremely hard on him, but we never shared our innermost emotional feelings about the possibilities of my dying. He was always so positive with me, saying, "You're different. You're not like all the others. Look, you've outdone the statistics already and you can do it!"

These were our main emotional adjustments along with accepting that my role was different from other mothers. They could care for their families quite well by themselves. I needed constant help from Jerry, our mothers, and other friends. It was hard to accept the fact that I needed this assistance.

Those were the main areas Jerry and I discussed when the possibility of having a kidney transplant came to us. Was it worth the gamble? Dr. Drake said my chances were about the same now whether I remained on the machine or had a transplant. Should we attempt it or remain where we were? Basically, Jerry left the final decision up to me. He would go along with whatever I decided.

After much prayer and soul searching, the answer came to me. Yes, I would go ahead and try a transplant. The opportunity was before me, God had been good to us so far, and I would follow His guidance into a new adventure. If it proved successful, Jerry and I would have a whole new life and would truly be living again. It was worth the gamble!

I called Dr. Drake the next day and told him to go ahead and make that appointment with Dr. Lawson at the University of Oregon Health Sciences Center. I would go talk to him, see if he would accept me on the kidney waiting list, and give it a try. After all, God was in control of my life.

Those who sow tears shall reap joy. Yes, they go out weeping, carrying seed for sowing, and return singing, carrying their sheaves.

—*Psalm 126:5-6*

Chapter 10

Kidney at Last?

It was a clear, crisp fall morning when we left for the University of Oregon Health Sciences Center. It was only a test, I was told over and over again. Sometimes the center would not accept patients for transplant. Your body had to be in good condition and be a good risk before they would even consider transplantation. The bladder, in particular, had to be in good shape. When it is not used for quite a while, it shrinks down to a very small size and a cystoscophy has to be performed to see if it is still usable. If it isn't, transplantation is often not the answer.

Then, there were the blood tests and the tissue typing. My friend, who had been on the machine about as long as I, had gone to the center to inquire about a kidney transplant, but had been turned down. Her blood was a very unusual type and chances of finding a kidney to match her were very remote.

What if they found something out about my blood that would make a transplant unrealistic for me? I thought, "Put your faith in the Lord and He will sustain you."

My negative thoughts left quickly as this Bible verse came to me. I knew God would choose the best path for me to follow and all I had to do was to listen and to trust in Him.

Today would be the initial testing. I knew they would just talk to me, take a medical history, get blood samples, and arrange for a time when I could be admitted to the hospital for what is called a "work up."

This is when the cystoscophies are performed. The doctors look with a cystoscope up into the bladder and the old kidneys, if you still have them, which I did. They also do heart tests to see how strong your heart is and essentially just test your entire body to see if it is suitable for a transplant. At this time, they like to know if a patient has diabetes, which poses additional risk.

Jerry pulled the car up by the round circle parking area near the front door to the outpatient building.

"Why don't you get out here," he said. "Then I'll go park the car."

"Okay," I replied. "I'll wait for you inside."

My mother worked at the center in the ophthalmology department and I knew from experience that parking was a real problem.

Once inside the Center, we were overcome by the massiveness of the building and didn't know where to go. We inquired at the information desk.

"Can you please tell us where the urology clinic is?" we asked.

"Take the elevator that's straight down this hall and then on the left. It is on the fourth floor," the receptionist politely said.

As we walked down the long hallway, (I was sup-
ported by only a cane now and my husband's strong
arm) we noted the people along the way. I will never
forget the mother pushing her little crippled child
in a wheelchair. Then there was the man with only
one leg, and the teenager struggling down the hall,
whose body was crippled by muscular dystrophy.
How fortunate I felt. Here I was, walking with leg
braces and the help of a simple cane! Yes, I
remembered the lonesome days of people staring
at me in a wheelchair and my thanks once again
went up to my heavenly Father. He had been so good
to us—despite the situations we were living through.
There were people who truly were less fortunate
than I and my position was not to ask why, but to
thank God and glorify His name for what He had
given to me.

We reached the elevator now and pushed the button for the fourth floor. Then the wait and much
anticipation set in.

What would they say; what would they do? As
I thought, all of a sudden a fearful feeling like the
pounding of a huge kettledrum emerged inside of
me.

"Ding," the elevator bell sounded.

We stepped onto the elevator—again with people
who were obviously less fortunate than we. There
were little children dressed in ragged clothes and
mothers who looked desperate for a better way.
Then there was the teenage girl, heavy with child,
and the old man whose crippled frame was bent
over bones that almost seemed to protrude from the
very skin that held him together.

"Cancer," I thought and I turned to look the other
way.

"Yes, we certainly are lucky," I thought. God

had restored our faith by giving us a perfectly healthy little son who brought constant joy to our lives. He had also brought us through some very precarious circumstances and another test in faith was before us. Things could have been much worse.

I glanced up at the numbers lighting up above the door on the elevator. Everyone was silent as we reached the fourth floor. The elevator went no further and everyone got off here. We stepped out into the unknown and ahead of us was another information desk.

"Can you please tell us where the urology department is?" Jerry asked.

"Do you have an appointment?" the girl inquired.

"Yes, with Dr. Lawson."

"Okay, straight down this hall and above a door about halfway down, you will see a black sign that says 'Urology.' They will help you there."

We soon saw the black sign and a nurse greeted us at the door.

"May I help you?" she asked.

"My name is Ann Van Winkle and we are here to see Dr. Lawson," I stated.

"Oh, yes, Ann, he's expecting you. Won't you please have a seat out in our waiting room?"

She pointed to an area located down the hall just a few feet. It was a room decorated nicely but conservatively in blue and beige. Chairs were lined up arm to arm and situated in long rows with people sitting back to back.

"Have all of these people had transplants?" I thought to myself. I soon discovered this was a waiting area for other clinics located on down the hallway.

After thumbing through many magazines, we were stirred by a voice.

"Ann Van Winkle," the lady called.

"Yes, right here," I replied as I slowly arose from the blue vinyl chair that had encased my body.

"Need some help?" Jerry asked.

"No, I'm fine," I replied.

"Dr. Lawson will see you now," the woman said. "Please follow me."

She placed us in an examining room and it wasn't long before Dr. Lawson entered. He was a prominent surgeon at the Center who had worked closely with Dr. Hodges, the doctor who had done the first transplant "on the hill," as they called this area of town. Dr. Hodges had transplanted the kidney from one twin to her ailing twin sister in 1958. These girls became the first transplants done at the Health Center and it was very successful. (At this writing, these women continue to lead healthy, full, energetic lives.)

Dr. Lawson went on to do many more kidney transplants and it was exciting to learn of the progress that was being made in this area.

"Hello," he said. "I'm Dr. Lawson."

"We're glad to meet you," Jerry replied. "I'm Jerry Van Winkle and this is my wife, Ann."

"I have your medical records here, Ann," he said, "but, I'd like you to go over your medical history for me, if you wouldn't mind."

So, I began to relate to him the experiences I had gone through: the strep throats I had as a child; the three pregnancies my body had undergone within the span of one and one-half years; the loss of the first two children and our very successful third birth with Jeff. I continued by telling Dr. Lawson all about the kidney infections that began occurring six weeks after Jeff was born and the trips to various doctors before our favorable encounter with Dr. Drake, who along with others, got me on the kidney machine and saved my life. I related to him the story of having my legs fail, of learning to walk again, the days of removing two-thirds of my stomach as a result of bleeding ulcers, and the problems I had with calcium, the parathyroid operation, and my experience with a tracheotomy. He asked me several questions and then outlined their plan for me for the next few months of my life.

"We would like to do some blood and tissue typing today," he said. "Also, assuming these tests come back favorably, we would like to set up a time when you could come into the hospital for a work up. This would be a three- to four-day stay for you and at that time, we would be able to do more extensive tests."

"All right," I replied. "That sounds good to me. The only question I have is would you run me on the machine during the work up time or do you want me to rearrange my home schedule so I could have three days free in the middle of the week?"

"No, we can run you," he said. "It would be good for you to get used to our machines. Sometimes we even have to run transplant patients until their kidneys function."

"Is that right?" I questioned. "How often does that occur?"

"Well, not often, but you know sometimes the kidneys don't work right away. They go into shock and sometimes don't work for five to ten days or more."

"That's interesting," I replied. "I guess I thought that once you had the kidney that the only problem was the rejection factor. I've heard of people rejecting their kidneys even after a few weeks."

"Yes," that's true," he replied. "But, we have been

having a good success rate. With cadaver transplants such as your prospective one, we have about a fifty percent success rate. Now, do you have any more questions?"

"I have a few notes written down here that I wanted to be sure and ask you. Just let me see if we've covered most of them. One thing I wanted to know is how long would I have to wait before trying a transplant? That is, if all the tests go right?"

"We never know," he answered. "With cadaver transplants it depends upon availability of kidneys. You may wait a few weeks, a few months or maybe even a year. We take people according to need so we really can't promise anything."

"I understand," I said. "I'm doing fine on the machine, so right now I'm not in a rush, but anytime would be fine."

"We'll do our best," he added.

After those final comments, he directed us to the nurse who would set up an appointment slip to have blood work done at the lab. She would also arrange a time for my admittance to the hospital for a work up.

When the appointments were all arranged, we left the Center with assurance that in the future, a new experience awaited us and it was both scary and exciting!

The blood tests came back favorably and it wasn't long before I found myself entering the university's hospital—this time for the long-awaited work up to see what condition my body was in. Could it stand the trauma of a transplant or would I be better off living the rest of my life on dialysis?

Nurses, doctors, technicians and auxiliary personnel at the Health Sciences Center flocked to my room. The question remained: could my body be matched with someone else's body? Someone had to pass away and will their kidneys to a waiting recipient. It had to be a match close enough so I could have new life through the gift of the donor's kidney.

Blood technicians came to do more extensive blood studies. Medical school students, interns, and residents all gathered around my bed or at the door to talk about the work up and the secrets their tests were telling about my condition. The peripheral neuropathy that was in my legs and hands was of much interest to all.

I shall never forget the morning some students and a resident were to examine my bladder to see if it could hold urine again. The nurses prepared me for the cystoscophy procedure and I was wheeled upstairs to an examining surgery room. This would be the real test. My blood and tissue typing had passed and now the decision of transplantation lay in whether the bladder was acceptable.

The resident and students helping were very down-to-earth people and we developed an instant rapport. They knew how important their discoveries were to me, but they tried very hard to make light of the situation. Upon examination and finding out that my bladder was acceptable, one fellow exclaimed: "It's the most beautiful bladder I've ever seen. It's gorgeous! Can you imagine its delight to once again be swallowed up in wonderful urine?"

"Let me take a look!" another fellow begged. "Oh, my land! It is gorgeous. Now that would knock the socks off a rooster. Out of sight! I know lots of folks who would love to have a bladder just like that!"

"Come on, fellas. Are you pulling my leg? What do you really see?" I pleaded as I lay spread-eagle on the examining table while their eyeballs peered through the cystoscope up into my bladder.

"No, it's really true!" one said. "It's great. Bladder-wise, you won't have any trouble."

"Whew!" What a relief I felt. Maybe it really was going to come true. Maybe, just maybe, I would really be able to have that long-awaited kidney transplant. However, the thought was somewhat frightening. I realized that the road ahead was not smooth. There would be chuckholes and ruts that would have to be conquered one at a time.

Another test in my work up had me somewhat concerned. It was the one that would test my heart. This was important because the heart goes through a lot of stress conditions during transplant. If it wasn't strong enough and failed to beat, then it was all over. There had been some concern earlier in my dialysis years about the possibility of my heart having a slight murmur. However, no sign of this appeared on these tests, so everything was "GO."

The doctors released me the next day to go back home. It was back to my own kidney machine and back into the routine of daily living while we awaited a call from the university. How long would it be before a kidney would be available for me?

My friend Susanna, who lived across the street, had introduced me to another neighbor named Jean. Jean was a tall, stately, attractive blond and a nurse who worked on the kidney transplant team at the Center. Discovering that she lived just down the street from us was really surprising but reassuring. Here was a personal source that I could go to and ask question after question about transplants. She advised me it was worth the chance, based on the many successful transplants she had seen. I'm sure she minimized talking about the unsuccessful ones and about the ones whose kidneys were rejected, but I knew that was a side to the story also.

Jean and I spent many hours over tea and ice cubes discussing the transplant situation. It was a few months later that the telephone rang out through the stillness of a crisp winter morning.

"Mrs. Van Winkle?" the voice asked.

"Yes," I replied.

"This is the urology department at the Health Sciences Center."

"Oh, yes?" I replied again, getting excited.

"We have your name on our transplant waiting list and we just might have a kidney available for you today. Are you still interested?"

"Am I still interested?" I thought. "Is this really true? Could they really be calling me?" The last several years flashed through my mind like a movie projector film going backward. I stood almost in shock, when I again heard the voice on the other end of the line.

"Mrs. Van Winkle? Are you there?" the voice questioned again.

"Yes, I'm here," I said. "I'm sorry, I guess I'm so excited my mind drifted away for a minute. I've waited a long time for this phone call and just can't believe you are calling me!"

"I understand," she answered. "Are you still interested in having the transplant?"

"Yes, of course. It sounds great!" I answered.

"Now, please understand this is only a speculative call. We have the cadaver, but will need to wait a few more hours before we can obtain the kidneys. I will call you back about one o'clock to let you know if we can proceed with the transplant," she said.

"All right. Sure. I understand. You just wanted to alert me to a 'maybe' situation," I said.

"That's right," she answered. "It does look like

everything is 'go' at this moment, but circumstances could change. Now, just in case we can do the surgery tonight, please refrain from eating or drinking anything until I talk to you at one o'clock."

"Sure," I answered. "That won't be any problem. I'm too excited to eat anything right now anyway."

"Good, I'll call you at one o'clock then. Good-bye," and I heard the phone go "click" and the line was still.

Slowly I hung up the receiver and stood by the kitchen counter wondering what to do next. My first thought was to call Jerry or maybe I should go over ot his office and tell him.

"No," I thought. "I'd better stay by the phone. I'll just call him."

My next thought was to call every prayer chain in the world so they could be praying for me.

Now, just wait a minute, I thought. Maybe I should keep this under my hat. God knows what He is doing. My life is in His control and maybe I shouldn't tell everyone in town just yet.

"Go shout it from the rooftops," one side of me was saying. But, the other side was saying, "Are you sure you want to go through with this? You know that you're doing pretty well on the machine right now. You know some of the stories you have heard about transplants. It's not going to be a bed of roses. In fact, you might just be jumping from the frying pan into the fire. Are you sure you want to go through with this? Maybe you should call them back and just say that the whole deal is off. No one would even have to know."

But I suddenly snapped back to reality and found myself dialing Jerry's office number.

"Is Jerry there?" I questioned.

"Just a moment, please. I believe he is on another

line. Would you like to hold?" she asked.

"Sure, I'll hold," I replied. "This is his wife."

"Oh, hi, Ann. I'm sure he won't be long," she replied.

The wait, however, seemed the longest I had ever gone through when all of a sudden a voice said, "Jerry Van Winkle, may I help you?"

"Hi! It's me," I replied. "Guess what?"

"What?"

"The university just called and they might have a kidney for me tonight."

"Great!" he answered. "What did they say?"

"They have the cadaver right now, but they have to wait a few more hours to see that there are no more signs of life. Then if there aren't, I'll go in this afternoon and have the transplant tonight. I just can't believe it and now I'm getting cold feet!"

"Cold feet? How come?" my husband asked.

"I guess I'm just scared," I answered.

"It will be okay," Jerry assured me. "Do you want me to come home?"

"No, I'm okay. It's just that we've waited such a long time for this and now maybe it's really here. It's just kind of scary."

"I know, but you'll be just fine," my husband reassured. "Leave it all to 'the man upstairs.' Okay? Now, call me back as soon as you hear anything."

"Okay," I said, "I will."

With those parting remarks, we hung up the phone with our standard "UDDO," which meant, "I do love you, honey."

"Call Jean," was my next thought and I automatically dialed her number. She would be able to answer all of my questions.

As I dialed I was saying a silent prayer that she would be home. I really needed her right now. From her experience on the transplant team at the Medical School, she would know the answers, or where to find them. There were a million and two questions I had on my mind.

"Ring--ring--ring," the telephone rang so deliberately slowly. "Hurry up," I thought. "Please be home."

"Hello," a voice answered and I knew right away it was Jean.

"Hi, Jean. This is Ann," my voice rang with excitement.

"Well, hi! How are you?" she answered.

"Very excited at the moment," I replied. "The Medical School just called and they may have a kidney for me today!"

"That's great! What can I do for you?"

"Well, how about coming up for a cup of tea and some chit chat. I'm as nervous as a new bride and just need someone to talk to," I answered.

"Okay. I'll be right up. Travis is asleep and I'll see if Loree can come over and stay with him awhile. Everything's going to be just fine so try to stay calm. You have to be in good shape in case this works out for tonight."

"Okay," I answered. "I'd come down, but the nurse is going to call me back and I think I'd better stay by the phone."

"Yes, you'd better. I'll be up in just a few minutes," she replied.

As I hung up the receiver I thought of a dozen people I could call.

"Maybe I'd better wait a while, though," I thought. "Better wait until they call me back and not get everyone excited just yet."

I washed a few dishes that were in the sink and busied myself around the kitchen while I waited for Jean.

"Knock, knock," a voice called in my front door.

"Come in," I answered. It was Jean appearing as calm and collected as always.

"How are you doing?" she asked.

"Pretty good considering this may be my chance of a lifetime," I answered. "Come on in the kitchen and I'll fix you a cup of tea. Just maybe in a few days I'll be drinking a cup of tea with you. Wouldn't that be great?

"Well, what do you think?" I questioned. "Shall I go ahead with it if this is for real? Do you think my chances are pretty good?"

"Well, you know it's a risk," Jean answered. "But don't you think the risk is worth it? I know it won't be easy, but, provided everything goes okay, you'll feel so much better. I've seen a lot of successful transplant cases and most people feel just great about their new kidney."

"Oh, just keep telling me that," I pleaded. "You're the biggest encouragement I have. I know you've been right there and have seen this surgery done a hundred times or more. Tell me again how it's done!"

As the clock slowly ticked on, Jean again went over the whole procedure with me. She had been Dr. Lawson's head surgical nurse and related every step to me. The only reason she left the program was because she was pregnant. I was sure with her self-assurance and patience she had been a terrific nurse. She was getting anxious to get back into the field.

As she finished telling me about the transplant procedure, I could see an idea illuminating in her eyes.

"You've got something up your sleeve," I said. "What is it?"

"How would you like for me to ask Dr. Lawson if I could scrub for your surgery tonight? That is, if they call you back," she questioned.

"Oh, that would be perfect," I said. "Do you think they'd let you?"

"Won't hurt to ask."

"Perfect," I said again. "How about another cup of tea?"

"Sounds great."

Just when I finished pouring her a cup of hot, spiced tea, the phone rang. My heart did a flip-flop as I answered.

"Van Winkles," I said.

"Good morning, Mrs. Van Winkle. My name is Sheri. How are you today?"

"Fine," I answered.

"That's good," she answered. "Mrs. Van Winkle, we have a special offer on our all-weather, aluminum siding during this month. We have a free gift to give you in exchange for just a few minutes of your time. Could I set up a time when our salesman could come out and show you our siding. Please understand there is no obligation to buy."

I couldn't get a word in edgewise, but I certainly was not interested in aluminum siding today. Kidneys, yes! Aluminum siding, no! I stood on one foot and then on the other as I tried to listen politely to her sales pitch. She finally finished.

"I'm sorry," I told her. "I'm not interested today."

"May I call you on another day?" she persisted.

"No, thank you," I replied and hung up the phone.

"Can you believe that?" I said to Jean. "Aluminum siding on a day like this! That's the least of my worries today. I'll bet I could have blown that girl's mind if I had told her about the phone call I was standing here waiting for."

We both laughed. It was good to laugh and we both agreed that the humor of the whole situation was good for us.

"Ring--ring," went the phone again. I quickly picked it up. Maybe this was the university!

"Van Winkles," I said.

"Hi, have you heard anything?" a voice asked.

It was Jerry, and the anticipation was getting the best of him so he just had to call.

"Not a word, honey," I answered. "I'll call you as soon as I hear anything. Jean is here with me and we're just doing transplants all over the place!"

"Okay," he replied. "Call me when you hear something."

"Will do. They should be calling pretty soon. Talk to you later."

I barely hung up the phone when it rang again.

"Van Winkles."

"Ann?" a voice questioned.

"Yes," I replied.

"This is Dawn at the Medical School. It looks like we have good news for you. If everything goes all right, we'll be doing your transplant tonight."

"Really?" I answered. "I can't believe it! That's fantastic. Now what do I do?"

"Come to the Medical School about three or three-thirty. Check in and they will direct you to our floor. Just bring some personal items with you and we'll take care of the rest. Oh, and remember, don't eat or drink anything," she directed.

"Okay, now just one question," I said. "Do you know what time the surgery will begin?"

"Probably around seven p.m., if all goes well. Now, don't worry. Just leave everything up to us," Dawn calmly said. "We'll see you around three. Bye now."

With that I hung up the phone and turned to Jean.

"Well, it looks like it's a 'go.' I'm so nervous I can't even think straight. What do I do next?" I asked.

"Call Jerry first," Jean said.

"Yes," I thought, "then there's my mom, Jerry's mom, my grandma, the Bacons across the street, my minister..." and on and on the list grew. It wasn't long after my phone call to Jerry that he appeared through the back door and all I wanted was for him to hold me. We had been through so much together and this might be the beginning of a new life!

Jeff, who had been at school during the traumatic hours of waiting and wondering, was just about due home and I knew Jerry could tell him better than I could about what was going to happen. He always had a way of talking with Jeff.

During all the commotion, Jean had called the Medical School to talk with Dr. Lawson about allowing her to be present during my surgery. He said it would be fine and this made me feel quite secure knowing she would be by my side. Everything seemed to be going in our favor.

Two-thirty, the time to leave for the hospital, rolled around rapidly. After packing my suitcase, talking with Jeff and making all the necessary phone calls, we left for the hospital. Jean would meet us there about five p.m. after all of the preliminary tests had been done. This was fine with me because Jerry would be with me meanwhile.

Checking into the hospital was routine procedure, and then we were directed to 5C, the transplant ward. Nurses, interns, and residents greeted us and tried to make me feel comfortable in this new environment. Preparations were done for surgery and it wasn't long before Dr. Lawson, who was to perform the operation, entered my room.

"Hello, Ann," he said. "It looks like everything is progressing smoothly. If our timing is right, we will be taking you up to surgery about six-thirty p.m. The operation is scheduled to last about four to five hours, but it can go longer or sometimes a shorter length of time. We are doing two transplants tonight. You understand we transplant only one kidney into you, so the other cadaver kidney will go to a young man from Coquille, Oregon. We will put the kidney in a cavity located under your hip bone and hook it up to the bladder from there. Now, do you have any questions?" he asked.

"No, I guess not," I said. "Only the impossible ones such as, 'Is it going to work?' and I know you don't know the answer to that one. So, I'm just trusting you with my life."

"We'll do the very best we can," he said. "Everything should go just fine. Now, Jerry, will you be waiting here in the hospital during the surgery or will you be going home?"

"Probably, if it's okay with Ann, I'll go home and get our son to bed. Jean will be with Ann and I don't think I'd be much good around here after she's gone to surgery," he explained.

"All right. Then I'll contact you by phone when we're finished. We have your number at the desk, I'm sure. I'll see you in surgery, Ann," he said and left the room. His positive attitude remained.

Jean came at five o'clock and it wasn't long until six-thirty arrived and they were wheeling me off to surgery. I was filled with much apprehension and joy as Jerry kissed me good-bye, saying he would be back in the morning. He walked me as far down the hall as he could and Jean accompanied me the rest of the way to the surgery room. Yes, this was a new, unknown adventure, but I knew in my heart that I was in God's plan.

Everyone was pleasant in the cold, sterile surgery room and there was a feeling of excitement in the air. I learned that they were using three surgery rooms and two sets of transplant teams for this operation. I was in one surgery room, the cadaver that was donating the kidneys in the middle room, and the recipient of the other kidney in the third room!

At this very moment the anesthesiologist was inserting a clamped-off line of Sodium Pentothal into my right arm, and a team of doctors was in the middle operating room removing the cadaver's kidneys.

The nurses were in a state of suspension awaiting news to go ahead and put me to sleep and start preparations for the operation. It wouldn't be long now before the kidneys would be removed from the cadaver and transplantation procedures would begin. Time was important. Dr. Lawson did not like to have the cadaver's kidneys out of the body any longer than necessary. Sometimes this caused them to go into shock and they were not apt to work as rapidly.

A nurse entered from the cadaver surgery room followed closely by Dr. Lawson. He came to my side and I could tell immediately that something was wrong.

"Ann, we don't have good news," he said. "We're going to let you go back home. The kidneys don't look good and I'm not going to chance the transplant. I know how important this is to you and I know we can find some healthier kidneys for you."

"You're the doctor," I said. "I understand and I certainly respect your opinion!"

We bid each other farewell and he promised that my name would remain at the top of their waiting list. When a good match for my body came in, I would be the first to get it.

Jean, who had faithfully stood by my side during the dream that didn't come true, helped to reverse all medical procedures to get me off the operating table, back down to my room and back home. What a godsend she was!

I decided not to call Jerry from the hospital, but to ride home with Jean. She lived only two houses down the street from us and I suppose the saga and the surprise of it all to Jerry intrigued me. And that he was—SURPRISED. Here I was, back home only one-and-one-half hours after he had left me at the hospital. There was a lot of explaining to do, a lot of phone calls to make, and a lot of trusting to learn.

Yes, God is in control of our lives and we don't know why he puts us through the tests, but He does. Waiting was the next test to be learned and we practiced its lessons daily. Down deep in my heart, however, I felt an unexplainable relief that this transplant attempt had failed. I knew that out there in the future God had control and what He had for my life was ultimately good.

Chapter 11

Second Chance

Christmas was in the air as we got back to the
daily routines after the "almost transplant." It
had been a dramatic soap-opera-type of situation to
go through, and I was glad to busy my mind and
my days with the coming of the holiday season.
There was always much to do. Besides buying the
gifts and preparing the house, I continued teaching
my twelve piano students. I loved teaching. It was
something I could do at home so I would be there
when Jeff arrived home from school. I was so glad
that my hand muscles had improved enough so I
was able to play the piano again.

This year I was planning to have a Christmas piano
recital at home, which would occupy a good share
of my time. I loved baking the cookies to serve, finding just the right punch recipe, and seeing that the
children carefully prepared their piano pieces. The
busy preparations made the time go by quickly, and
before we knew it, we were into the new year.
It was 1973 and the year-and-a-half that followed

progressed with minimal excitement. At the beginning of 1974, the Medical School called to assure me that they had not forgotten about my much desired kidney. I remained on their transplant waiting list. They would call just as soon as a kidney became available.

Dr. Drake and Dr. Hayes continued to support me in the role of dialysis that we pursued weekly. Jerry still ran me on the machine at home three nights a week. The biggest change that occurred during this time was a new dialyzer that became available through Drake-Willock Company. Because of this new, more efficient method of cleansing the blood, I only had to run eight hours, three times a week on the machine. This meant I didn't have to go on dialysis until nine-thirty p.m. and would come off at five-thirty a.m. What a difference this made in our evening life! I had two more hours to do the dinner dishes, to get Jeff to bed, and maybe spend some time playing the piano before I had to get on the kidney machine. This wasn't bad and I settled down to really accepting kidney machine life. Maybe tranplantation wasn't for me after all. I was doing just fine on the machine and maybe I would stay right here.

As many life situations evolve, just when you settle down to really accepting what *you* think is God's plan for your life, a change is imminent. My case was no different.

Jerry and I often went for walks in the balmy spring evenings that were upon us. It was good exercise for the body and particularly my legs that were still encased by leg braces. I discovered that the stronger I became, the more I was able to do for myself. It was a good feeling.

We had returned this particular evening from a walk to the state park that was located near our home in West Linn. It was a challenge for me now to walk clear to the farthest end of the park which was down a steep hill that ended at the river. We would often tarry here, feeding the ducks and enjoying the peacefulness of the river before walking back up the steep hill and be homeward bound. We returned about nine-thirty p.m. this particular evening, and, as usual, after such a long walk, I was tired. This was not a kidney machine night, so, after getting Jeff to bed, Jerry and I collapsed into our king-size bed, anticipating a peaceful night's sleep. It was then ten p.m. and in the stillness of the night the telephone rang.

"Who in the world is calling us at this time of night?" I wondered aloud.

Because the telephone was on my side of the bed, I reached to answer it.

"Van Winkles," I answered.

"May I speak to Ann Van Winkle?" the voice inquired.

"This is Ann," I replied.

My heart flip-flopped for something inside told me this was not a salesperson calling. This call had an air of importance surrounding the voice on the other end of the line. Who was it? My instantaneous thoughts were swallowed up when a voice answered again.

"Ann, this is Mary Anne at the University of Oregon Health Sciences Center. How are you this evening?"

"I'm just fine," I answered. "I'm a little tired, but just fine."

"Well, Ann, we may have a kidney for you. A possible cadaver donor has been located and it appears that the blood and tissue typing matches your body. Are you interested in trying again for a transplant?"

"Sure, sounds great! What do I do now?"

"Actually," she informed me, "there's nothing you can do but try to get a good night's sleep. We won't know until morning if the kidneys are good. We will take them out of the cadaver then and put them on our new pump. This will keep a solution pumped through them and keep them functioning until we can transplant them. Now, as I said, the best thing for you to do is to get a good night's sleep and I'll call you in the morning. We just wanted to alert you to this possibility so you wouldn't be leaving town."

"Okay," I replied. "I'll do the best I can."

By this time, Jerry knew, through listening to my end of the conversation, what the phone call was about. He had laid down his nighttime book and was waiting for me to fill him in on all the details.

We talked for an hour or more before we finally turned out the light to follow doctor's orders.

"Get a good night's sleep," she had ordered, but how in the world was I ever going to sleep with this exciting news coming right at bedtime or at any time for that matter! This phone call could be the first step in changing our entire lives.

"Proceed with caution," my mind was echoing in the background. "Remember what happened last time. Don't get too excited. This may be another false alarm.

"It might be," I thought, "but I don't think so. We'll see in the morning," and I turned the light off and settled down in prayer to thank God and to turn the situation over to Him. My body was tired after the long walk to the park. It was not long before I drifted off to sleep.

The spring morning appeared quite early through the crack in the blue flowered bedroom drapes. I awoke and suddenly remembered the phone call of the night before and realized that this may be the day of all days. May 28, 1974, could be a date to remember for the rest of my life. The anticipation of the hours and days that lay ahead were almost overwhelming.

I crept out of bed as quietly as I could. Jerry was still sleeping and it would be good for him to enjoy this peacefulness a bit longer. I had learned to walk a little bit without using my leg braces. By holding onto the windowsill, the dresser, and then the door knob, I could get out of the bedroom and work my way down the hall to the family room. I could sit in there and meditate until it was time for Jeff and Jerry to awaken.

I just reached the bedroom door when Jerry stirred.

"Where are you going?" he asked.

"Just to the family room," I said. "Go back to sleep for a half hour. It's not time to get up. You might have a long day ahead of you."

With this, I slipped out and closed the door. It was not long before Jeff was up and busy getting himself ready for school. He was in the third grade now and a very energetic boy. Jerry was ready to leave for work when he called Jeff aside and explained the phone call of the night before to him. We wanted him to understand that Grandma might be here when he got home. I could be at the Health Sciences Center being prepared for a kidney transplant.

I'm sure he did not comprehend the magnitude of the situation because we didn't even fully comprehend it ourselves. We all did know, however, that if this transplant worked, there would be no more kidney machine and I would probably feel much better. Could this really happen? This morning would bring the first answer to a multiple-question situation.

Jeff and Jerry both kissed me good-bye.

"Be sure to call me at work, honey, as soon as you hear something," he said.

"Don't worry," I replied, "I will. Now you have a good morning and don't worry about me. I'll talk to you later. Drive carefully!"

I stood in the driveway as they both left, Jeff walking to the school bus with his friend, Gardner, and Jerry driving off to work.

I decided to read the morning paper before I got dressed, and had just settled down when the phone rang.

"Golly," I said to myself, "could it be the Health Center already?"

"Van Winkles," I answered.

"Hi, honey. How are you this morning?"

It was my mother and I didn't want to tell her about the possible transplant until I knew for sure. So, I answered her questions pretty casually.

"Oh, I'm fine. I'm just sitting here reading the paper and playing 'lady' this morning. It's a beautiful day."

"Yes, it is up here, too. I can look out of my office window over toward Mt. Hood and it is a beautiful sky this morning," she said.

My mother worked at the Health Science Center where I might be going for my transplant this very day. If this whole situation worked out, it would be convenient for her to stop by and visit me on her lunch break. That would be comforting.

We passed the time of day for a few more minutes and hung up the phone saying we'd talk to each other later. I thought to myself," It may be sooner than she thinks. If the Center calls with the go ahead,

Mom would be one of the first to know." My dad would have been excited, too, but he had passed away a few years before.

I busied myself after talking with Mom by making a list of people I would be sure to call if the Health Center called me to come in. When my list was just about complete, the phone rang.

This could be it, I thought.

"Van Winkles," I answered.

"Ann?" a voice questioned.

"Yes," I replied.

"This is Mary Anne at the Health Center. It looks like a transplant is in order for you today. We have obtained the kidneys from the cadaver and have them on our kidney pump right now. They look good and we'll be transplanting both of them today. One will go to a girl from Medford and one to you. Now, can you check into the hospital by two p.m.?" she asked.

"Sure," I replied. "We'll be there. This is an exciting day. I can't believe it! Can you tell me where the kidneys came from?" I questioned.

"Well, I can't," she responded, "but the doctors may be able to fill you in on more of the details. We'll see you at two p.m. Try to stay reasonably calm and don't eat or drink anything."

"Okay," I replied. "See you at two."

I hung up the phone and could hardly believe this was happening—again. It was almost a rerun of one-and-one-half years ago when the kidneys turned out to be defective. I prayed this would not happen again.

I proceeded to call most of the people on my list— Jerry being the first and my mom second. They were all excited. Jerry said he would be home soon to take me to Portland. Just after I called my mother, she called back to say that Mary Anne had just brought my new kidney by her office to show her. The organ was on a kidney pump and was encased in a plastic-covered box.

"It looks real good," she said. "I know you're going to get along just fine."

No sooner had I showered, dressed and packed my suitcase when my husband walked through the door.

"Anybody home?" he called. "Where's my bride?"

"I'm in here," I answered.

He held me and kissed me and somehow I knew this time everything was going to be all right. A new life for us was about to emerge and it was exciting!

It was soon time to leave for the Health Center. We made sure that Jeff would be taken care of when he arrived home from school and we were on our way.

We arrived in plenty of time. I was checked in at the main desk and taken to my room on 5C. I was asked to undress and put on a hospital gown, and then preliminary tests began.

Blood pressure and temperature were taken. A technician from the lab came to take blood for some extensive tests, and interns, residents, and other doctors paraded to my room taking my medical history and explaining what lay ahead. They explained that my operation would be the second transplant of the day. The woman from Medford was being transplanted at this very moment and I would go to surgery about six p.m.

"Try to get some rest," I was told. "You have a lot ahead of you."

As the hours slipped by, Jerry stayed by my side

and word spread throughout the hospital that the kidney being transplanted into the Medford woman worked as soon as they hooked it up.

"How lucky," I said. "I hope I am as fortunate!"

I did remember something that one of my transplant friends had told me. Katheryn's kidney had not worked for about ten days and she said not to give up if mine didn't work right away.

"Don't give up, no matter what," she had said. "It will work. It has to!"

Before I knew it, I was being prepared to go up to the surgery room and I felt like I was replaying a similar scene.

Jerry kissed me good-bye and said he would be back in the morning. Jeff needed him at home and the doctor would be in touch with him by phone.

"Everything is going to be just fine. You just go in there and show them how to do it! Okay?"

"Okay," I replied. "Give Jeff a kiss for me. 'Night, honey."

With that farewell, they rolled me down the hall and into the elevator up to the surgery floor.

The double-wide surgery doors were in front of us as my escort rolled my gurney out of the elevator. After checking in with the surgery desk, we entered the well-equipped surgery room.

"Hello, Ann," a masked nurse remarked. "We're going to try it again, I see. This time it's for real."

"Were you on my last fiasco when the kidneys weren't any good?" I questioned.

"Yes, but this time is going to be different," she assured me.

"I certainly hope so," I answered.

"Now, can you scoot yourself over onto the surgery table?"

As I scooted with my arms and laid back onto the

hard, narrow surgery table, she said, "Might as well get comfortable. You'll be here a while, but of course you won't know the difference. In fact, here's the anesthesiologist to talk to you."

"Hello, Mrs. Van Winkle. I'll be giving you a little stick with a needle in just a few minutes and then I'll ask you to count backward from ten," he explained.

"Fine," I replied, "I think I can handle that."

No sooner had the nurse taken my blood pressure and inspected the site near my left hip bone for the incision when the anesthetic needle was stuck into my vein.

"God," I thought, "it's all up to you now. I give my life to you. Guide the surgeon's hands. Whatever your will is, let it be done."

"Count backward from ten," the anesthesiologist said.

"10, 9, 8, 7, 6...," I muttered and I was asleep and the surgery began. New life and new hope was within hours' reach, and somehow I felt at peace. God was in control.

•   •   •

When I woke up hours later in my hospital room, I felt sore, very drowsy, and in a rather confused state of mind.

"Did it work?" I questioned the nurse who was standing by my bedside.

"Your operation was very successful, Ann. You need lots of sleep, so just lie back and don't worry about anything," she comforted me.

"Did it work, though?" I asked. "Did the kidney work?"

I was remembering that the woman from Medford who received the other kidney had the joy of her kidney working right on the operating table. I wanted to know if mine worked also.

"No, it isn't working yet, but it will. Just give it a chance. Transplanted kidneys are often like this. It will work," the nurse said. "You just lie back and get some sleep."

"Sleep," I thought. "Yes, I am very tired. Sleep sounds good. Real good. I'll ask more questions in the morning."

"Mrs. Van Winkle," I heard a voice call out by my bedside. "Mrs. Van Winkle, I need to take some blood from you."

"Okay," I muttered holding out my left arm. "Here's my cannula. Just take it from there. My cannula clamps are around my neck."

Taking blood for testing had been a routine procedure when I was on the kidney machine. Just unwrap the bandages around my arm and the cannula, take off the tapes holding the two tube ends together, clamp off the tubing and pull the tubing apart gently. This exposed a small sylastic sleeve tubing that was put into the blood vial and upon releasing the clamps just a little bit, blood would seep out and fill the test tube. So easy and so painless! I would hate to go back to being stuck with needles. But I knew there would be no need for this cannula if my new kidney worked, because there would be no more need to run on an artificial kidney machine!

"By the way," I drowsily asked the technician. "Do you know if my kidney is working?"

"I don't know," she answered. "The doctors will be making rounds pretty soon. You can ask them. Thanks for the blood," she said as she finished wrapping my cannula back up. "Try to get some more sleep now. It's real early."

I laid back and had no trouble going back to sleep.

I felt real dopey from the surgery and I wanted to sleep more than anything else right now.

I was again awakened when Dr. Lawson and some medical students entered my room.

"Good morning, Ann," he said. "Your surgery went just fine. We have you on Prednisone and Imuran medication to help keep your body from rejecting the kidney. The kidney hasn't started to function yet, but this is not uncommon. We look for it to start working any time. We have a catheter in you which will enable us to monitor how much urine you will be producing.

"We are also going to start you on a series of what we call ALS shots. These are rather painful but we believe this helps in the antirejection of transplanted organs. You will be getting one shot a day for fifty days—that is, if you don't show any reaction to them. If you do, we will take you off of them immediately.

"Just one more thing," Dr. Lawson continued. "We will leave the cannula in your arm for a few days. If your kidney does not function right away, we will be running you on the kidney machine and we will need access to your blood."

"That's okay," I replied. "'This cannula and I have become pretty good buddies. I've had it now for seven years."

Most of the first three days were spent either in sleep or asking if my kidney was working. I remember Jerry being by my bedside for many hours during those days. His careful monitoring of my catheter and the bag that hung on the bedside by it always produced the same answer.

"No, not yet," he would say. "Just give it a chance, though. It's going to work." My faithful friend and nurse, Berdie, would come by to give me encouragement, but the answers were always the same.

The next recollection I have is when Dr. Lawson told me they would be dialyzing me on the fourth day. My blood was being filled with uremic poisons which needed to be filtered out. It was back to the kidney machine, and I felt plenty of discouragment.

"Don't give up," I would tell myself. "God knows what He is doing—at least I think He knows." My thoughts, emotions, and physical body were so confused at this point that I really didn't know if I was in full control. In fact, everything got so bad that I told Dr. Lawson I thought I was going out of my mind. It felt like I was on the outside looking down at myself. It was an eerie feeling.

He assured me that I probably felt like that because of the high dose of steroids that I was on. They couldn't lower the dose too rapidly, or my body could reject the transplant. He said that they would lower the dose as rapidly as possible. I was satisfied with this. Just knowing it was the medicine and not me made a big difference in my thinking.

Days four, five, and six slipped by and still no function from my new kidney. I was moved into the room with my "kidney sister," as we called each other (the woman who received the other kidney from the same cadaver). Her kidney continued to work beautifully and I had to repress my envy daily. It was hard being in the same room with her.

Days went by and I was run on the kidney machine again and continued receiving my ALS shots. They hurt and we all complained about them.

One man down the hall who was receiving an ALS shot had a cardiac arrest during the middle of the shot. They brought him out of it, but the rest of us were told that the heart problem had nothing to do with the ALS shot. We really didn't believe them; however, we continued taking the shots, only because it seemed like we had to. If we quit and then rejected the kidney, we would have always been sorry.

Days seven, eight, nine, and ten went by and still no function from my new organ. I was scheduled for the X-ray scanning room where I lay motionless for an hour or more while the technician set a machine to scan my new kidney. They wanted to see if there was a reason it was not functioning. The results of the tests were inconclusive. The kidney just wasn't working yet. Sometimes the kidney goes into shock from being on the kidney pump too long, I was told.

It was back onto the kidney machine and much more waiting ahead for us. Jerry continued to be encouraging and neither of us lost hope. We just kept praying, "Please, God, let it work!"

Days eleven, twelve, thirteen, and fourteen were much like the days before and then came day fifteen—the day that was to change my life.

A nurse entered my room and with much enthusiasm shouted, "IT'S WORKING, IT'S WORKING!"

"Oh, glorious day—PRAISE THE LORD," I shouted. "It's really going to work," and my bedside kidney sister and I rejoiced together. Now there were many phone calls to make and Jerry was the first.

My mom was excited, Jerry's mom was excited, my friends, the hospital staff—everyone flocked to see. Yes, it was a glorious day. I called Dr. Drake to tell him the good news and we rejoiced together.

Twenty cc.'s of urine the first day, fifty cc.'s the second, and I shall never forget reaching five hundred cc.'s one grand and triumphant day.

"There's no place to go but up now," I thought. "Everything will be just fine now." God really was in charge and it was good.

Chapter 12

A New Life

"*P*izza!" I exclaimed. "Pizza is what I want!" This "no-no" during the seven years I was on the artificial kidney machine suddenly sounded delicious. "Pizza would taste so good," I told my husband. "Just think, if this kidney really takes off, I'll probably be able to eat anything and drink anything," I said excitedly.

"What do you mean 'if' it takes off?" he questioned. "There's no doubt in my mind that it's taking off right now. Just look at your jug. It's one-third full of urine!"

I looked over at the counter in my hospital room where my gallon jug sat. Each transplant patient's urine was measured very carefully and recorded on a chart posted on our door. On the same chart the patient's intake of fluid was recorded. By comparing the two figures, the doctors could tell right away how well the new kidney was working. If the output was not as great as the fluid input, the kidney was not functioning well. However, so far, mine

149

was running "neck and neck" with my input.

"Drink more," the nurses would say. "Give your kidney something to do. We'll be taking the catheter out soon and then you're on your own. We'll see how well your bladder is working then."

"Any day now," I thought, "Dr. Lawson will let me eat anything I want to eat." My mouth once again watered just thinking about the hot, spicy, cheese and Canadian bacon pizza I desired. I couldn't think of anything better.

"I'll tell you what," my husband said. "As soon as Dr. Lawson gives the 'go ahead,' I'll go get pizza and we'll have a pizza party right here in your room! How would you like that?"

"Sounds fantastic," I said, and I reached for another glass of water. The nurse told me that the glass-and-a-half I had the day before was not enough. I needed to drink eight or ten glasses of water a day. What an order, but I tried my best to comply. This was a new learning experience to me. After not drinking anything for seven years, I really had to learn to drink all over again. This new life was going to be superb and I loved every glorious minute of it!

Always, however, in the furthermost corner of my mind was the thought of organ rejection. I was fully aware that the body does not like foreign objects in it. It was explained to me like this: When you accidentally get a sliver in your finger and don't take it out right away, the finger gets infected around the sliver and tries to push it out by itself. This is called rejection. The defenses of the body know that this is a foreign object and they "gang up" to remove it.

Rejection of kidneys works basically the same. Several people I knew had rejected their kidneys soon after the transplant. Others had rejected their new organs as many as two to nine years after transplant. I knew full well that this was a possibility, but I never liked to think about it.

As I reached for another glass of water, I picked up the mirror that was lying on my bedside table. Glancing in it gave me an uncomfortable feeling. My face was becoming rounder and puffier every day. The area under my chin and jawbone seemed to be at least twice its normal size and I was breaking out in a horrible rash.

"It will go away," the doctors assured me. "We have you on a high dose of steroids right now so that you are not so apt to reject your new kidney."

They were talking about the Imuran and Prednisone medicine I took daily. For some reason, the Prednisone made the body retain more fluid in given areas such as the face and neck, the back of the neck just above the spine, the abdominal area, and around the tops of the knees. The two hardest areas to accept these changes in were the face and abdomen. A puffy face made me feel like I was someone else when I looked into the mirror. And the distended abdomen made me feel like I was four to five months pregnant.

I tried to fully believe the doctors, however, because I wanted everything to turn out for the best.

My kidney sister and I began building a good relationship during our rooming together days. We felt so good about our new kidneys working that we became known as Queen I and Queen II and respective signs were hung at the foot of our beds. We drew many comments from doctors, residents, and interns as they made their daily rounds. Carefully they would go over each patient and determine the next necessary step in their total care.

Sometimes it made me feel like I was just another number being reviewed daily for teaching purposes, but eventually we learned that this was not true. After building a rapport with the nurses and doctors, I felt like they really did care about me.

Eventually the team of doctors decided that my catheter would be taken out.

So, upon receiving orders, out came the catheter. I was told I would have to make many trips to the bathroom now since the bladder would be collecting the urine. As soon as it felt full, I would need to go. Since it had not been used for seven years, it had decreased in size and could not hold much. However, it would soon stretch back, the doctors constantly assured me. They were correct. Soon after the catheter was removed, I felt pressure in my bladder. Praise the Lord, it was working!

From that day forward, my praise for God rang out loud and clear every time I went to the bathroom. This simple, taken-for-granted bodily function always reminds me that God cares about me and that He sent Jesus to die on the cross for me so that I could experience the abundant life. I felt I had that abundant life now.

As my organ began functioning better and better, my daily diet began to improve. No longer did I have to be careful about what I ate or drank. For the many years I was on the kidney machine, I longed to drink a bottle of pop. This became reality one day when the nurse asked if I would like some 7-Up.

"Can I really have one?" I asked. "I mean, is it really okay?"

"Sure," she said. "Just go help yourself to one in the refrigerator across the hall."

This I hastened to do although I felt that I was doing something wrong. My mind and body had been programmed against drinking for such a long time that it was difficult to overcome this long, ingrained way of survival.

About twenty-five days after my transplant and ten days after the kidney started working, my husband popped his head around the corner of the door and shouted, "SURPRISE!"

His arms were laden with pizzas for all! What fun we had. My kidney sister, Pam, and I could hardly wait to taste the pizzas that smelled so good! We invited some nurses, doctors, and other patients to come and join us, and we had a celebration party right there in our room. Again, a small wish had come true and this new life that I was experiencing was getting more desirable every day.

Everything was not perfect, however. Problems started to develop that were difficult to cope with. When Dr. Lawson arrived in my room one morning, he greeted me in his usual professional manner.

"Good morning, Ann. From looking at your chart here on the door, it looks like you're doing pretty well. How do you feel?" he asked.

"I feel pretty good," I answered, "but sometimes I feel kind of funny."

"How's that?"

"It's hard to explain," I answered. "But I get very confused in my mind. In fact, it sometimes seems like I'm outside my body floating around in the air looking down at myself. It's very frightening and sometimes I think I'm going out of my mind."

"How often do you feel this way?" he inquired.

"Most of the time, I'd say. It seems like the only escape I have from it is when I go to sleep. I've been trying to sleep a lot lately, but I know that's not a cure—just an escape. Do you know what's causing it?"

"I suspect it's the Prednisone," he answered. "We

will lower your dose as much as possible but the only problem is, we can't lower it too fast. Your body would be too apt to reject the kidney and neither of us wants that!

"Now, what about this low-grade fever I see recorded on your chart?" he questioned.

"That has just developed," I answered. "Is it true that this could be a sign of rejection?"

"It could be," Dr. Lawson responded. "I don't think it is yet, though. Your blood tests have been coming back from the lab indicating your chemistries are improving. We'll wait and see about the fever. It may go away."

The fever did become a real mystery and stayed with me for many weeks to come. After eight weeks in the hospital, I was finally discharged, still carrying a low-grade fever. When I went home, I developed a slight infection around a hangnail on my left thumb. Pus began forming around it, and after a culture, the doctors discovered it was a staph infection! This was a word that always elicited fear from the transplant doctors. Staph infections and low body resistance did not go together at all! Immediately they put me on an antibiotic and soon the thumb infection and mysterious fevers subsided. I will always believe that the staph infection was hybernating somewhere in my body and God chose to reveal it by a tiny infection in my thumbnail.

Before I had been discharged from the hospital to go home and continue my newfound life, the cannulas in my left arm had to be removed. The last segments of this old way of life had to go, and in a way, it was hard to give up the cannula. Even though it was no longer needed, this tubing in my artery and vein had become a part of me. Carefully I had cleansed it daily and wrapped it with a sterile gauze and an Ace bandage to keep it clean.

The surgery procedure turned out perfectly. It was not long before living without a cannula became an enjoyable part of my new life. Now I could take a bath or a shower and actually get my arm wet. There was no fear of infection setting into my arm because there were no open areas where germs could crawl in and set up "home." I was free—really free! It was truly a "born again" feeling. This new life was great and once again I praised God.

Life at home after my kidney transplant was the beginning of a whole new ball game. Newspaper people, friends, acquaintances, and family were eager to see how the game was played, and I felt a big responsibility. I wanted them to see that God was playing the game through me and that He had the control.

It was a marvelous feeling to be free of the kidney machine that still stood in the corner of our bedroom. In a way, however, I was glad that it was there when I arrived home. Like my cannula, it had been a real living part of me so it was like a security blanket to see it still standing there in case I needed it. Losing a transplanted kidney did not mean one had to die. It meant that either you got another transplant or went back on a kidney machine.

The machine remained in our bedroom for about two weeks. Then Chuck Foster of the Kidney Association of Oregon called me.

"Hi, Ann. This is Chuck," he said.

"Well, hello," I replied. "Good to hear from you."

"How are you feeling?" he asked.

"Pretty good," I replied. "Having a kidney of my own is great and I'm not having much dificulty at all adjusting to this new way of life. It's great! I can eat or drink anything I want now."

"That's wonderful," he replied. "I don't suppose you'd like to get rid of an old friend of yours, would you?"

"Well, I don't think I'll argue with you about that one. My machine has been like an old friend, and, in a way, it's going to be hard to let it go. It's like it has been a part of me—you know?"

"Well, I can imagine what it's like, but guess I don't actually know. You and that old machine have been through a lot of ordeals together. We'd like to pick it up tomorrow or the next day. Would that be all right with you?" he asked.

"Sure. So far I haven't ventured too far from home."

"Great. We'll see you probably tomorrow morning. We'll give you a call before we come out."

"Okay. Good to talk to you, Chuck," I ended.

As the days passed by, I began to feel better and better. I had forgotten how it felt to really feel good—I mean to feel so good that you actually had energy to stay up late at night and even felt fantastic when your feet hit the floor in the morning. No more naps in the middle of the day so I could teach my piano students and still have enough energy to fix dinner and get on the kidney machine.

Feeling good—that was the name of *this* new game. Everything about life was great right now. Jerry would tease me about my new kidney coming from someone who was a late night go-go dancer because I loved to stay up late. As my energy level improved, I began doing many other activities I had not had the chance or energy to do for years.

Jeff was in the fourth grade by then, and it was God's timing that I would be more available to him and his needs. I loved becoming involved in school and church activities with friends and family. Life became greater than I had ever imagined.

Chapter 13

And Tomorrow Came!

"**A**re you hurt?" Jerry called out as one of my skis slipped out from under me and I took a tumble in the soft snow. He came to a graceful but abrupt halt right by my side.

"I'm okay," I replied. "A small chunk of my pride fell off—that's all!" I shook my head as snow sprayed my already cold, frosty face that peered from beneath my cockeyed ski hat. There I lay, twisted and a bit shaken in the cold icy snowbank on the side of Mt. Hood. But it was a good feeling. My dream of being able to ski had come true.

With help from Eddie in the brace shop at the Medical School, my newly-designed leg braces allowed me to venture upon this mountain slope. Encasing my lifeless lower legs, these new braces could fit snugly into a pair of ski boots.

"Come on, Mom. Come on and try," Jeff had said. At fourteen years of age, he was urging me to participate in his favorite sport with him and his friends.

Why not, I thought. I'll never know if I can do it unless I try. So, I tried and I did it! Success was the prize again.

Traci, Emily, Paula, and Rick, all teenage friends, were always nearby to give encouragement. Jerry again was patient and taught me the skills of the sport, since he was an experienced skier.

The first day on the lift at Mt. Bachelor was an unforgettable experience. Making it onto the chair, getting off at the top, and then the slow, cautious trip on my own down the mountain brought unexplainable joy. I could really do it. Leg braces and all, I was skiing!

Tomorrow had finally come and I loved every minute of this new adventure: going to the grocery store and buying anything to eat, dining in restaurants, staying up late at night, and just having a family again whose life evolved around family activities instead of a kidney machine. It was great!

One of the most exciting events that occurred was our trip to Hawaii. Now I could be away from the kidney machine for more than a few days. Boarding the airplane with our good friends the Kohls, the Sparks, and the Bacons was better than a fantasy. This was reality and a new, joyful life.

Along with skiing and bicycle riding, I found energy to do many other outside activities. I became very active in our church and took charge of the Sunday school department for two years. Jerry's manufacturers' representative business with the Don Raymond Company, Inc., had grown and I became his personal secretary.

Music has always been my "first love." Besides continuing to teach my sixteen piano students, I took jazz piano lessons for myself (another dream realized) with the help of my teacher, Chloris Alexander.

One of the most gratifying experiences I have had since my transplant has been fundraising. Organizing benefits for the Kidney Association of Oregon and the Renal Transplant Research Department under the competent direction of Dr. John Barry at the Medical School has allowed me the privilege of giving of myself to two organizations that have given so much to me.

I thank God when I think about the many Christmases I have seen since those days on the kidney machine. I thank Him also for allowing me to see my child grow up. Jeff, at seventeen years of age, is growing into a fine young adult.

My kidney is now nine years old. How can I ever thank its anonymous donor enough? A tragic death that brought life could come only from God through the family of the generous donor.

Through all of our years of marriage, Jerry and I can look back and say, "Through it all, life has been good. We praise the Lord."

For right now, it's full speed ahead and smooth water. We set out to make the most of every minute **of every** hour of every day. God is the pilot of our **ship, and where** He directs, we will follow. My new **life is in** His hands and I commit it to Him daily.

> After you have suffered a little while, our God, who is full of kindness through Christ, will give you his eternal glory. He personally will come and pick you up, and set you firmly in place and make you stronger than ever. To him be all power over all things, forever and ever. Amen.
> —2 Peter 5:10-11